THE BATTLE OF BUFFALO WALLOW

The Japanese Attack on the 44th General Hospital In World War II – Leyte, Philippines December 1944

James R. Odrowski

Seven Cedars Press
Lenexa, Kansas

Copyright © 2020 James R. Odrowski
Seven Cedars Press
Lenexa, Kansas

ISBN: 978-0-578-78674-2

Cover: 44th General Hospital photo with doctors, nurses, officers and enlisted men. Lt. Edward Odrowski (author's Dad) is in the middle, third row from bottom, fourth man from the left. Lt. Walter Teague is on the same row, second man from the left. Colonels Weston and Waddell, first row from the left.

Contents

Acknowledgements

Thanks to those who made this book possible, particularly the veterans who took the time to tell their stories, and the many historians who've painstakingly documented the context of the vast and complex conflict known as World War II.

Special thanks to my wonderful wife Colleen who put up with the long hours spent by me on this project. She is truly inspiring; nothing can match the time I spend with her. I look forward to returning to the trout streams of Wyoming with her, very soon.

My admiration goes out to Lt. Col. Walter A. Teague, U.S. Army Medical Administrative Corps (MAC) and Captain Eda A. Teague, U.S. Army Nurse Corps (ANC) who served with the 44th General Hospital. After their retirement, they worked diligently to document the story of their unit. They also spoke at schools, sharing their stories of serving in a military medical unit with many young students. Walter, an avid photographer, returned with over 300 photos that captured the people and places described in this book. The Teagues' archive is currently housed at the Wisconsin Veteran's Museum in Madison, WI. It includes memories from other 44th General Hospital veterans, including Drs. Chet Gjertson, Ray LaFauci, Jackman Pyre, Edward Birge, James Bingham, and Herb Pohle. The 44th's nurses who told their stories included Eda Teague, Ida Bechtold and Emily LaDuke.

Chet Gjertson and Walter and Eda Teague also left recorded interviews at the Library of Congress Veterans History Project. They can be found online at http://www.loc.gov/vets/. Also, thanks to Colleen Janes, writing about her Dad, Richard C. Janes, who was a supply sergeant with the 44th. She shared her Dad's stories on a blog site, https://potrackrose.wordpress.com/2013/09/20/guest-post-battle-of-buffalo-wallow-wwii/.

I heard my Dad mention the names of fellow veterans in his stories. Although I didn't get to meet them in person, I feel like I got to know them all. Their humility and sense of humor came through in their memoirs. My Dad, also an avid photographer, brought back over 200 photos from the War. Some of these are provided in this book. They provide a unique glimpse of the people, places and events he experienced during the War. Thanks to my sister Rita who preserved many of Dad's photos.

I appreciate the veterans who wrote about their experiences in the Pacific Theatre, particularly those who served in the same campaigns as my Dad. They provided valuable insights into the events they shared. These include Donald O. Dencker, who served in the U.S. Army's 96th Infantry Division. Private Dencker described the events he experienced on Leyte in his book *Love Company: Infantry Combat Against the Japanese, World War II*. Also Dr. George Sharpe, who provided a combat physician's view of the war in the Philippines, in his excellent book *Brothers Beyond Blood*. Sally Hitchcock Pullman provided a nurse's viewpoint in her book called *Letters Home: Memoirs of One Army Nurse in the Southwest Pacific in World War II*. Sally followed the same path as the 44th, and was assigned to the unit at the end of the War.

Another very sobering account of the War was provided by two courageous women of the Philippines. They recounted their painful memories of sexual abuse by the Japanese military as "comfort women." Maria Rosa Henson told her story in *Comfort Woman: Slave of Destiny*. Remedios Felias, with the help of Chieko Takemi, wrote and illustrated her similar memories in *The Hidden Battle of Leyte: The Picture Diary of a Girl taken by the Japanese Military*. Their work brought to light the atrocities committed against them and over 200,000 other young female victims during the War.

I greatly appreciate the historians and authors who helped me connect the historical context to my Dad's stories. Their painstaking research was of great value. The key authors and their works included Nathan N. Prefer's very comprehensive *Leyte 1944: The Soldier's Battle*, Clayton Chun's *Leyte 1944: Return to the Philippines*, G. Rottman's and A. Takizawa's *Japanese Paratroop Forces of World War II*, and Gene Eric Salecker's *Blossoming Silk Against the Rising Sun: U.S. and Japanese Paratroopers at War in the Pacific in WWII*.

Special thanks to The National Museum of the Pacific War, located in Fredericksburg, Texas, and the Wisconsin Veterans Museum located in Madison, Wisconsin, for their excellent exhibits and collections.

A warning to the reader. I've included graphic depictions of the realities of the War. Also, when recounting the stories of the veterans, I wanted to reflect their language and feelings of the times. They had been engaged for over four years in a brutal battle with a very determined enemy. I mean no disrespect towards the present-day Japanese people when quoting the veterans. The Japanese people have been great allies and economic partners since the War. My son, currently serving in the U.S. Navy, spent a number of years stationed in Japan. The citizens have been very warm and welcoming to him.

Last, but not least, I appreciate the time my Dad took to tell me his stories of the War. Although he enjoyed his children, grandchildren, hobbies, and business career, it was evident from his stories that the War was the greatest adventure of his life. I appreciate that my Mom also provided her perspective of the times, describing her life as a newly-wed Army wife. Her stories helped me understand what it was like on the home front, caring for a young son while helping her parents run their corner grocery store. She experienced three tense and anxious years, waiting while Dad and both of her brothers served in the Pacific Theatre. Also, thanks to my friends, Larry, Terry, and K.C., who enjoyed listening to Dad's stories as much as he liked telling them. Dad appreciated the opportunity to share the experience of the War with our generation.

Walter Teague, reminiscing about his and his wife Eda's service with the 44th General Hospital stated,

> It is with the 44th that we share common memories about a group of very uncommon people. It was an honor to be with such splendid men and women during the most memorable part of our service. We are sure it was a never-to-be-forgotten time to all who served with us. Fearing the truth of Caesar's law that, "the evil that men do lives after them, the good is often interred with their bones," we hope that an accurate and official record of the 44th's service in

the Pacific will be preserved. In this, we are all challenged by these lines from the Iliad, "Now, though numberless fates of death beset us which no mortal can escape or avoid, let us go forward together, and either we shall give honor to one another, or another to us."

For my Dad, Walter, Eda, and the rest of those who served in the 44th General Hospital, what follows is my effort to tell your story, to be the *other* that gives honor to you for your service and courage.

Cheers!

James R. Odrowski
September 2, 2020, V-J Day
The 75th Anniversary of the end of World War II

Preface

I've read that for the most part the veterans of World War II "didn't want to talk about *it*." Many felt that if you weren't there, you wouldn't understand. Some were tormented by the memories of what they saw and what they had to do. Some were racked with guilt, asking why they lived and their buddies didn't. Some just wanted to get on with their lives, feeling fortunate to have made it back alive.

It was different with my Dad. For him, it must've been therapeutic to talk about it. He only told his stories to a few of us who were eager to listen. I soaked in his stories of the War from an early age. The stories ran the gamut from suspenseful to comical, to graphic and disturbing. It seemed that the War had been the greatest adventure of his life. But he didn't glorify war or wish the experience on anyone. Many times, he'd end a story by bluntly acknowledging that "war is hell" (quoting Civil War General William Tecumseh Sherman).

The black and white photos he brought back implanted visual images in my mind of the people, places, and events he experienced. There were also the artifacts that stirred my young imagination. His helmet, uniforms, an M1 rifle, bayonets, and the colorful battle ribbons. Three items intrigued me the most. They included a Japanese sword, a Filipino bolo knife, and a strange insignia with a pair of coiled snakes. The insignia was a caduceus, a symbol of the Army Medical Corps. These three items represent the defining event described in this book, The Battle of Buffalo Wallow.

I didn't grasp the historical significance of his stories at an early age. Many questions stirred in my young mind. What was it like to be there? We're you shot at? Did you shoot at anyone? Did you see people die? How did you get through it knowing that you might not make it back home? Later in life my curiosity and interest in military history got the best of me. Many years after my Dad's passing, I was driven by the need to "connect the dots." I sought to link historical context to the stories he told. I had no reason to think that he embellished anything, but still pondered the reality of what I heard. As I found out through my later research, the actual events were even more remarkable than his stories implied.

This book tells the story of the 44th General Hospital, a dedicated and courageous U.S. Army Medical unit. My Dad served in the 44th as an officer of the Medical Administrative Corps (MAC). The 44th trained at Ft. Sill, OK, before deploying overseas to Australia, New Guinea, and the Philippines. They took part in the Leyte invasion and the battle for the Philippines, a pivotal turning point in the Pacific War. The 44th was made up of "citizen soldiers." It was staffed by experienced doctors and nurses from the University of Wisconsin-Madison. Admirably, the medical professionals put their practices on hold and risked their own safety to serve others in the war effort.

General Norman T. Kirk, Surgeon General of the Army during World War II, called the 44th General Hospital "the finest that ever served." Kirk praised the actions of the 44th in *Collier's* magazine, dated July 22, 1945. In the article titled "That They May Live!," he described how they risked their lives in service to their patients. Finding themselves at "ground-zero" of the Japanese counterattack on Leyte, they were asked to perform actions over and above their medical duties.

As I complete this book during the 2020 COVID-19 pandemic, I'm reminded of the recent service provided by our front-line medical providers. They are also risking their own health and safety that "others may live." They serve as an example for us all.

The story of the 44th General Hospital is not without controversy. Medical units in World War II were supposed to be protected from direct attack. The Geneva Convention states that "hospital units should be inviolable" in war zones. Also, that medical personnel should not take up arms, so as not to be misconstrued as combatants. Japan had signed the 1929 Convention, but did not ratify it. The actions taken by both the Japanese and the 44th General Hospital in December of 1944 walked a fine line. In the heat of battle, life or death decisions had to be made. But when researching the story of the 44th many questions came up. Why was an Army General Hospital left vulnerable to Japanese attack? Particularly when U.S. intelligence intercepted enemy communications planning such an attack. Were the 44th and other supporting service units left to fend for themselves in light of larger military objectives? And a key question, why aren't they mentioned in official military records or in present-day historical accounts? Some of the veterans of the 44th believed that their story was conveniently covered up. Obvious gaps in documentation seem to support that belief.

Perhaps the story of the 44th General Hospital has just been lost in the many accounts of heroism during World War II. From the Pacific theatre alone, 464 United States military personnel received the Medal of Honor, 266 of them posthumously. The courage and commitment of those who received the nation's highest award is undeniable. So, by comparison, what is the significance of the group of men and women that served in the 44th General Hospital? Should their story just blend into all the others of great sacrifice and courage? I've researched the 44th's history over the past five years. In that time, I discovered that their service was even more commendable than this group of veteran's stories indicated. They faced significant risks that were met by their unique skills and courage. They held a firm dedication and commitment to each other and the ones they cared for. I believe that they were motivated by healing, not harming; by love, not hate.

In 2020, in the midst of a pandemic, we observe the 75th Anniversary of the formal signing of the Japanese surrender. September 2, 1945 marked the end of World War II, the largest armed conflict the world has known. The veterans who served have mostly departed and the memory of them is fading through time. The 44th's experience is another story among many, that tell of the courage, dedication, and toughness of the "Greatest Generation." But I believe that the story of the 44th's unique heroism deserves to be told. I hope by telling their story, the members of the 44th can rightly take their place among the many heroes of World War II.

Introduction

Dad and I sat on the patio gazing out over the red brick planters into our suburban backyard. Having just finished mowing and weeding, we rested and admired the afternoon's work. The sweet aroma of Mom's petunias blended with the smell of the freshly cut lawn. Another hot summer day in Kansas was fading into the evening. A large maple tree shaded us from the setting sun. I broke the silence, hoping to trigger a story, "Did you ever get shot at by the Japanese?"

As expected, Dad started in, "It was December 6th, 1944. We were on high alert at our camp on Leyte. It was the 3rd anniversary of the attack on Pearl Harbor and we'd been warned that the Japs might be planning something big." He finished off a beer and continued, "You know the Japs didn't like to fight during the day. But it was a different story at night. You'd get your head blown off by a sniper if you lit up a cigarette. When we sat in our foxholes, we had to smoke in a coffee can."

The screen door that led out to patio opened, "Ed, are you still fighting that war?" Mom quipped as she came down the steps from the kitchen. She carried a cold Bud and a bottle of Coke on a wooden tray with pretzels, sliced cheese and salami.

Dad, acting perturbed that his story was interrupted, went on talking, as if ignoring her comment. "You know that after the surrender I was asked to go to Japan during the occupation? I would've been promoted to major and been in charge of a large city. Mom would've lived in a mansion and had servants." He lamented, "but she said no." (As part of the U.S. occupation of Japan after the War, General Douglas MacArthur recruited officers over 6 feet tall to be overseers of the major cities. He wanted them to tower over the Japanese in their positions of authority. At a lanky 6 foot 2 inches, Dad, a captain, fit the bill.)

Mom promptly defended her decision, "Ed, you know that we had just bombed the you-know-what out of them, and little Eddie was not very well."

Dad conceded again on this matter. He welcomed another cold beer on the warm Kansas evening. Mom returned to the house, not wanting to get caught up in another war story.

In 1957 Mom and Dad moved to the suburbs from the Polish neighborhood in Kansas City, Kansas. It was the year I was born. The suburbs were growing. Larger homes with big backyards, new schools, and shops. Low interest, VA loans helped finance the post-World War II suburban expansion. Even though it was just a 10-minute drive from their old neighborhood, their families practically disowned them for leaving. Mom and Dad both enjoyed doing yard work. Mom with her flower gardens and Dad with his trees and the thick Zoysia grass that carpeted the lawn.

Dad built the large red brick patio behind our suburban home. The patio would be the stage where he'd tell stories of the War. We'd sit in redwood chairs cushioned and arranged side-by-side. The chairs were connected by a small table in the middle with an umbrella holder. Mom filled the red brick planters with colorful arrangements of flowers. The smell of pungent marigolds and sweet petunias filled the air on summer evenings. We'd watch planes fly overhead, approaching the Kansas City airport in the distance. At night we'd set up a short-wave radio with a long wire antenna that would stretch out to Mom's clothesline. The crackling of static and foreign accents gave the impression of being somewhere exotic. Lingering humidity, a citronella tiki torch, and the occasional buzz of a mosquito gave the sense of being on a remote tropical island in the Pacific. With a balmy breeze and a dark sky streaked by the Milky Way, I imagined being on the deck of a ship crossing the vast Pacific Ocean. Sometimes we'd spot the Russian satellite Sputnik dashing across the sky. In the mid-1960s it was a subtle reminder of the latest threat to the free world. Dad said that there would never be another war like World War II. At the time I didn't grasp the significance of his comment. As I learned more about the Cold War and the War in Vietnam, I finally realized what he meant.

The War Begins

Dad never mentioned any girlfriends other than Mom, although he had been a bachelor until he was 27. Likewise, Mom never talked about any other boyfriends. She was four years younger than Dad. Dad described seeing Mom at the weekly polka dances they attended. He said that she was the prettiest girl in the neighborhood and very sweet. Mom described Dad as handsome and very self-assured.

Mom and Dad were first generation children, both born to Polish immigrant parents. Their parents settled in Kansas City, KS, at the turn of the century. Poles were recruited to work in the meat packing houses that lined the Kansas and Missouri river banks of Kansas City. They were taught in a Polish Catholic school. Lessons were in Polish first, then English. Mom said that the nuns would punish them if they were overheard speaking English on the playground. The immigrant families had setup a "little Poland" in KCK with their own church, school, and stores. Many expected to come to the U.S. to work, save some money, then return to the homeland someday. But with two world wars, the Great Depression, and the communist takeover of Poland, that someday never came. Instead, they would be asked to send their sons and daughters to fight and serve, and to possibly die in conflicts in their former homelands and beyond. Many immigrant families across the U.S. did the same. The freedom and opportunity offered by America, they found, had a high cost.

Dad had been attending a local college, studying business and accounting. But the Great Depression slowed down his college plans, as he had to work in the packing houses during the day and as a movie theatre usher at night. During that time, everyone had to pitch in to support their families. As a teenager during the Depression, Mom had to quit school to work in her parent's corner grocery store and help with household duties.

The theatre Dad worked at was called "The Granada." He worked there with two of his high school buddies, brothers whose father owned the theatre and a local funeral home. During the hot "Dust Bowl" summers of the Depression, the theatre would close down. His fellow usher friends had access to a car. The three young men would pool their meager savings, pack their camping gear, and head down the dusty roads, traveling throughout the western U.S. They would stop and work along the way, on farms, at tourist resorts, and at some of the Civilian Conservation Corps (CCC) camps that did work at the National Parks and Forests. President Franklin Roosevelt started the CCC program during the Depression as a way to provide millions of young men employment on environmental projects. The CCC camps would hire Dad and his friends for kitchen duty or enlist them as laborers for clearing trails. They would work for a week or so, saving up enough gas and food money to keep them moving on the road, as they traveled through Colorado, Utah, Arizona, Wyoming, and Idaho.

Mom had never travelled very far from Kansas City. One exception was when she was 15 years old. Her wealthy aunt took her by train to the Century of Progress International Exposition, also known as the Chicago World's Fair of 1933. Her father, Michael, had started his corner grocery store in 1921, leaving his job as a butcher in the packing house. Her mother Anna worked at the Armour Packing House's offices, cleaning, cooking, and working as a seamstress. She was an accomplished cook and baker. Mr. Armour, the owner, would stay in Kansas City with his family for extended periods. When Armour's personal chef became ill, the kitchen manager, in a panic, asked Anna to step in. Armour was so impressed, particularly by the pastries and pies she baked, that he requested that she cook for him whenever he came to town.

Mom said that things got tough for her family during the Great Depression. The packing houses cut back on production. The Dust Bowl had devastated many farms, and unemployment rose to record levels. Her parent's store struggled to stay open. In desperation, her father declared, "Why are we here? We could just as well starve in Poland." Mom said that their steamer trunks were piled in the living room and they were waiting for paperwork to return to Europe. But as they waited, things began to get better. FDR's New Deal restored public confidence and the family decided to stay in America. It was a good thing they did. On September 1, 1939,

Hitler invaded Poland. Two days later, France and Britain declared war on Germany. World War II had begun in Europe.

The U.S. remained neutral as public opinion was largely against getting involved in another European war. But the Polish roots of Mom and Dad's families ran deep and there was much worry for the safety of those living in Poland. Their parents had brothers and sisters living there. They anxiously watched the newsreels and read the papers describing the Nazi advance. Then, tragic news arrived from the homeland. Mom's aunts, uncles, and cousins had taken cover in a bomb shelter during the Nazi blitzkrieg. During an air raid, a Nazi bomb made a direct hit on their shelter. One of Mom's aunts was killed. Other family members were wounded, some carrying scars for the rest of their lives. At the family's village church in Tarnobrzeg, established in medieval times, the Nazis removed the church's large iron bells, melting them down for war material (my Grandfather returned in the 1960s and paid to replace them).

On September 27, 1940, the Axis powers were formed as Germany, Italy, and Japan become allies with the signing of the Tripartite Pact in Berlin. Japan, which aimed to dominate Asia and the Pacific, was at war with the Republic of China. By 1941, Germany had quickly conquered and controlled much of continental Europe.

Dad attended college in Kansas City while working at the main office of the H. D. Lee Company. He was pursuing a business and accounting degree. But he realized that the looming war would likely to put his plans on hold. He also knew that he had to act fast or potentially lose an opportunity he had been waiting for the most; he proposed to Mom. Their courtship was short. They were married on May 26, 1941, at the Polish church, St. Joseph's in Kansas City, KS. The reception was held at Mom's parents' home, which was connected to their grocery store. As with all Polish weddings, there were kegs of beer, vodka, whiskey, and a polka band. Traditional foods on the buffet included golabkis (baked cabbage rolls stuffed with ground beef and pork sausage and covered in tomato sauce), pierogis (Polish stuffed dough fried in pork fat), kapusta (a sautéed sauerkraut dish with vegetables), her mother's famous fried chicken, and smoked kielbasa (a.k.a., Polish sausage). The Polish wedding cake tradition was for each person to get a piece of cake, then choose between a shot of vodka or whiskey to toast the newlyweds. It didn't matter how old you were. If you were old enough to stand, you downed a shot. Mom and Dad danced the polkas with each other and many others. It was a great party; the thoughts of entering a World War were put on hold for the moment. Mom's brother Frank, who would later serve on a Navy destroyer in the Pacific War, captured the day on the latest 8mm color movie camera. Many years later when we viewed the movies, Mom would point out their friends at the wedding who were later sent overseas. She would always point out those that didn't return.

Dad had planned their honeymoon to be a road trip out West. Mom's family was against her traveling to what they perceived as an untamed land. They were afraid that the newlyweds would be attacked by hostile Indians or eaten by wild animals. They even offered them extra money to stay. But Dad was set on showing her the places where he had traveled during the Depression. After a couple of days on the road, Mom called back home to tell her folks that they were safe. As it turned out, family members were still partying at her parent's house. Evidently, the booze and food had not yet run out, and most didn't realize that she and Dad had left!

The honeymooners visited Rocky Mountain, Yellowstone, and Grand Teton National Parks. Many of the roads out west were narrow and rough. Snow still lined the roads in the higher elevations of the Rockies. The honeymooners stopped in Salt Lake City, touring the lake and the Mormon Temple. On their way back to Kansas, they stopped at Colorado Springs. The area was a popular destination for honeymooners. The Garden of the Gods, Royal Gorge, and Pike's Peak were popular attractions, as was the Will Rogers Memorial on Cheyenne Mountain.

Will Rogers was an American icon. His untimely death in a 1935 plane crash was greatly mourned by the country. His wit and wisdom shaped a generation who listened to him on their radios during the dark times of the Great Depression. The Will Rogers Shrine of the Sun is constructed of native pink granite quarried from Cheyenne Mountain. A large stone tower with a staircase led to views of the Rockies and vast prairie to the east. Spencer Penrose, who developed much in the Colorado Springs area, including the Broadmoor Hotel and the Cheyenne Mountain Zoo, dedicated the stone structure to Rogers, before being buried there himself.

Mom and Dad walked up the native stone steps of the memorial. Near the entrance sat an old, frail Native American man in buckskin and full eagle feather headdress. A young boy sat next to him reading a comic book. The old man appeared to be napping. Upon noticing the honeymooning couple approach, the old man leapt energetically to his feet, grabbing his small drum and leather-covered drumstick. With a steady drumbeat he closed his eyes and began to chant, moving in short circles as the couple approached.

He approached Mom and spoke to her, "You young bride?" He placed a headdress and white buckskin vest on her. He continued to chant. His grandson explained that it was a marriage blessing song. Dad captured the scene with his 8mm movie camera. He then spoke to Dad, "I like your hat." Dad was wearing a white, wide-brimmed hat with a black band. He swapped Dad's hat for a buffalo headdress. Not exactly matching his maroon sweater, tie, and white spats, the tall, thin "paleface" made quite a sight. The medicine man continued, "You look like good soldier." The drumbeat's rhythm increased as he chanted. Dad improvised a war dance as Mom took photos. The grandson, with a concerned look on his face, said that this was a blessing for safety in battle. The old man, now visibly tired, sat back down to rest. The boy explained that his grandfather had been both a warrior and healer. He had fought in the Red River War as a young Kiowa warrior. He later became a respected medicine man. Dad, not yet in the Army or part of a war, was not sure what the old man meant by his blessing. They tipped the grandson and departed for the long drive home across the hot Kansas prairie. The medicine man resumed his nap along the cool stone wall.

After returning from their honeymoon, the couple discussed their plans for the near future. With war looming and a low draft number Dad decided that it was best that he enlist in the Army. At the age of 27,

5

and with some college behind him, he figured that it would be better to enlist than waiting to be drafted. Dad was always eager to control his own destiny, so he determined that joining the Army Medical Corps might be his safest option. Combat infantry was expected to be on the front lines. They stood a good chance of being the first to take a bullet or the impact of a mortar shell. Dad figured that the Medical Corps staff stood a better chance of being positioned a safer distance from the front lines. As part of a support organization, he could leverage his organizational and accounting skills. Not that Dad ever backed down from a fight. Growing up on the tough streets of Kansas City, you had to learn to fight. He had even been stabbed through the thigh in a fight while working in the packing house.

Dad was inducted at Ft. Leavenworth, KS, on August 4, 1941. It was decided that Mom would live with her parents while he was in basic training. Boarding a train at Kansas City's Union Station, he left for basic training at Camp Grant. Located near Rockford, IL, Camp Grant was established in 1917 to train Army infantrymen for World War I. In February 1941, Camp Grant was re-activated as an induction and Army Medical Service training center. During World War II, over 100,000 medical personnel were trained at the camp. Dad noticed that many of the new recruits were much younger than he was, some just out of high school.

At the time, the U.S. was not yet involved in the War. American isolation was still favored by much of the public and many politicians of the time. Many felt that it wasn't worth sacrificing American lives to get involved in what was perceived as "the rest of the world's issues." Even with the imminent likelihood of entering of the War, the U.S. military was unprepared. The Great Depression had taken its toll on arms production and had greatly reduced funding. The U.S. armed forces had been greatly reduced in size. Just prior to World War II, the U.S. was not even considered in the top 20 world military powers, barely ahead of Bulgaria and behind Portugal. The Army's facilities, weapons, tactics, and uniforms were still of World War I vintage. The relatively small Army Air Corps was still flying in open cockpit planes.

Basic training consisted of physical conditioning, a lot of drill and marching, firearms training, preparation for gas warfare, endless cleaning, and kitchen duties, known as KP. Dad said that while on KP duty he made a key decision. As he was peeling a mountain of potatoes, with the mess sergeant breathing down his neck, he determined that he'd rather to be the one giving orders. Dad believed in and taught that persistence and determination were key to getting what you want in life. He completed basic training in late November of 1941 as private first class. From that time on, he quickly rose through the enlisted ranks and determined to apply for Officers Candidate School (OCS).

Dad returned to Kansas City on holiday leave, just in time for Thanksgiving Day, November 26, 1941. He'd have a couple of weeks off before reporting to his next assignment. His last Sunday on leave was December 7th. Every American, who was alive on that day, remembers their "Pearl Harbor moment." For the generation, it was the defining event that changed their lives. Similar to September 11th for current generations, they remembered where they were, who they were with, and what they were doing. Mom and Dad had attended church with their families and were settling in for a big Polish dinner at Mom's parents. Mom's brother Frank, already enlisted in the Navy, heard a bulletin come over the radio. It was announced that Pearl Harbor had been bombed by the Japanese. A large part of the American Naval fleet was destroyed in the surprise attack. Over 2,000 U.S. servicemen, mostly sailors, had been killed. Sixty-eight civilians also died. Americans turned to their leader, President Franklin D. Roosevelt, for guidance. The next day the family gathered somberly around their tall, wooden radio cabinet to hear FDR address Congress. He declared war on Japan, famously proclaiming December 7, 1941 as "a date which will live in infamy." Everyone knew that their lives were about to change in significant ways.

From the onset of the Pacific War, the Japanese moved quickly to create their empire across the entire western half of the Pacific Ocean and the Southeast Asian mainland. Japan's forces quickly toppled Wake Island, Guam, Hong Kong, the Philippines, and dozens of other locations in the central and southern Pacific and on the Asian mainland. Their goal was to seize and occupy resource-rich Southeast Asia, giving them access to the natural resources they lacked—iron, tin, coal, oil, rubber, and rice. This would allow them to create a protective perimeter stretching from the Aleutian Islands in the north, down through the central and southern Pacific and around into the Indian Ocean. Tokyo's propagandists called it "The Greater East Asia Co-Prosperity Sphere." In reality, it meant that every person and every resource within that sphere would serve the Emperor, or face elimination.

Following the declaration of war, Dad received notice to report back to his base immediately. On December 12, Dad boarded a Rock Island Railroad passenger train to Camp Cooke, located near Lompoc, CA. It was clear from the postcard he wrote to Mom, that he wished that he was driving instead:

From somewhere in Arizona, Thursday, 7 a.m.

Dearest Baby, this train ride is absolutely terrible. The only things that I'm able to see is the ground and the hills that surround the railroad tracks. It's nothing like traveling by auto. The train misses all interesting sights, but I guess all trains are alike. This is a first-class train too. You'll find out how it is when you make your trip here. The three fellas I'm with and I are playing bridge from morning to night. Love, Ed

Camp Cooke was approximately 50 miles northwest of Santa Barbara. The camp was at a beautiful location very close to the beach. After Dad obtained housing he sent for Mom. She made the long train trip from Kansas City to California. The couple lived in an Army trailer on a hillside overlooking the Pacific Ocean. They had some enjoyable times there. Mom's brother Frank, in training with the Navy, was also stationed in California. Frank, his wife Josephine, and young son Raymond visited Mom and Dad at Lompoc. They spent afternoons on the beach and took trips into Santa Barbara. Mom and her sister-in-law toured the old Spanish missions and enjoyed the vast fields of flowers grown in the cool, moist California valleys along the coast. The couples enjoyed the fresh fruits, nuts, and vegetables that were grown in the lush California valleys. Not knowing what lay ahead of them, the couples made the most of the time together before they had to part ways.

Dad quickly moved up the ranks while at Camp Cooke. He was promoted to corporal on January 10, 1942, and from corporal to sergeant on February 3, 1942.

In spite of the beauty and recreation enjoyed at Lompoc, there were also issues with being in wartime California. In spite of being over 5,000 miles from Tokyo, the western coast of the U.S. was under threat of Japanese attack. In February of 1942, a group of Japanese submarines snuck undetected to a position off of Ellwood Beach, near Santa Barbara. They lobbed 16 shells at a large oil storage facility causing minor damages. The event led to public hysteria in what was called "The Battle of Los Angeles." Anti-aircraft batteries along the coast fired over 1,000 rounds at what they thought was a Japanese air attack. Five accidental U.S. civilian deaths were attributed to the chaos that ensued. No evidence of enemy aircraft was ever found.

To the far north, the Aleutian Islands of Alaska had been invaded by the Japanese in June of 1942. It took U.S. troops until August of 1943 to remove them from their positions on the islands of Attu and Kiska. Oregon, of all places, also received much attention from the Japanese. In June of 1942 two attacks occurred, one by submarine and one by plane. A Japanese submarine made its way to the mouth of the Columbia River and surfaced near Fort Stevens, a decrepit Army base dating back to the Civil War. After midnight it fired 17 shells towards the fort. Not wanting to give away their position, the U.S. Commander wisely did not return fire. The plan worked, the fort was spared, but their baseball field took a direct hit and was severely damaged. In another attack, a Japanese floatplane dropped a pair of incendiary bombs in a forest near Brookings, Oregon. They had hoped to start a major forest fire, but favorable winds and quick response by fire crews limited the damage.

These would be the only attacks on a military base and by air on U.S. soil during World War II.

The Japanese had also developed over 300 long-range "balloon bombs" carrying anti-personnel and incendiary explosives. They were meant to attain an altitude of 30,000 feet and ride the jetstream over 5,000 miles to the U.S. mainland. Once over land, they would release their payloads. Most were shot down, but some were seen as far inland as Iowa and Michigan. They had little effect. But a balloon that landed in Oregon killed a pregnant woman and five children as they unknowingly came upon it lying on the ground. Their deaths were the only combat casualties on U.S. soil during the War.

Blackouts along the West Coast were mandatory due to the real threat of Japanese submarines and air strikes. Mom and other Army wives volunteered for Civil Defense duties on base. Government trainers instructed them in blackout procedures, firefighting, and first aid. Some of the wives also worked as "coast watchers." I later joked with Mom that she should have received a combat medal for her time in wartime California.

A darker part of their stay at Lompoc involved the Japanese internment. In fear of possible espionage and sabotage by Japanese living along the West Coast, the U.S. Government issued an order giving regional military commanders the authorization to detain anyone with Japanese ancestry. Over 120,000 Japanese-Americans would be displaced from their homes and sent to detention centers across the western U.S. Many Japanese lived in the Santa Barbara area, some prosperous business owners with beautiful homes. As they were boarded onto buses and trains, they were restricted to travel with little more than a suitcase, leaving personal belongings and their homes behind. Mom and Dad said that groups of U.S. servicemen and their wives would go on evening raids of abandoned homes and ask them to come along. They both refused adamantly. Art, jewelry, and other personal possessions were being looted. Years later, they both talked about how they detested the thought of participating in the looting. At the time, Dad made it clear to those who reported to him that they would be reprimanded if he found out that they took part.

Dad was accepted into Officer's Candidate School (OCS). At the end of 1942 he was transferred from Lompoc to Camp Barkeley, TX, located in the Texas Panhandle, north of Amarillo. The camp was a hot, dry, windy, and desolate place. Dad completed Medical Corps field training there. Mom had been a dutiful Army wife, following Dad across the U.S. while he prepared to go overseas. To save money, Mom pressed Dad's uniforms according to military specs. As Mom did my laundry growing up, I was the only kid in school wearing pressed blue jeans. They would have a sharp crease running up the front of the legs. Likewise, any collared shirt I wore would be stiffly-starched and completely free of wrinkles. They would have passed any Army inspection.

Dad graduated from OCS in December of 1942 at Camp Barkeley. He was commissioned a 2nd Lieutenant in the U.S. Army's Medical Administrative Corps (MAC). In February of 1943, Dad reported to Ft. Sill, OK. There he would meet up with the medical unit that he would be with for the duration of the War, the 44th General Hospital.

The 44th General Hospital was formally approved by the Surgeon General on November 27, 1940, following his invitation to the University of Wisconsin Medical School to form an "affiliated unit." The medical unit would be made up of civilian doctors and nurses from Wisconsin General Hospital in Madison, WI. It was realized that to completely man such a unit would greatly deplete the staff of the hospital and the faculty of the medical school. It was up to Dr. W. S. Middleton, Madison, Dean of the Medical School, Dr. J. W. Gale, Chief of Surgery, and Dr. Frank Weston, Chief of Medicine, to select the professional personnel for the hospital.

When Dr. Weston took command most of the staff had already voluntarily entered the Army. The doctors and nurses had been commissioned as officers and reported to duty with the 6th Service Command. During the period of May 1942 to January 1943, the hospital personnel, including nurses, served at numerous hospital posts in capacities comparable with the positions they would hold in the 44th. The hospital was activated as a complete unit at Ft. Sill on January 19, 1943. The 44th was a fully-equipped hospital with an accredited medical staff for surgery, medicine, radiology, dentistry, and psychiatry. These dedicated men and women of all ages, gave up their careers for the war effort. They answered the urgent need to serve and save lives.

Colonel Henry G. Waddell, experienced in Army field operations, was the overall Commanding Officer of the 44th. Lieutenant Colonel Frank L. Weston, a doctor, was named as Chief of Medicine. Dr. Joseph W. Gale, was named Chief of Surgery. Colonel Ida Bechtold, was named Chief of the Nursing Service. The ages of the 44th's doctors ranged from the late twenties to well into their 40s. Many were married and had children. This was in contrast to younger Army GIs who would serve in the infantry, many just over 18. The 44th numbered 643 total personnel, consisting of 545 men and 98 women. The breakdown of the roster included:

- 485 enlisted men, including medical corpsmen
- 98 nurses
- 48 medical doctors and dentists
- 10 administrative officers
- 2 chaplains

Dad was assigned to the 44th General Hospital as an administrative officer in charge of the enlisted men, including those responsible for securing the hospital. The hospital staff from Wisconsin and enlisted men from across the U.S. were sent to Ft. Sill for basic training. Dad's job was to train them in Army field operations in preparation for deploying overseas. The medical staff was given a crash course in soldiering, learning how to march, live in tents, set up facilities, manage supplies, and operate a hospital in a hostile, wartime environment.

The men and women of the 44th endured long hikes with heavy packs, and overnight bivouacs. They had to endure a tear gas drill. With tear gas enveloping them, they had to be able to put on a gas mask and continue breathing as they recited their name, rank, and serial number. They also had to complete an infiltration course where they crawled on their bellies under concertina wire, while live ammunition was fired over them. But, due to the insistence of high-level Army commanders, the medical staff and corpsmen of the 44th were not provided with arms training. The MAC officers had recommended that they should be trained, but they were told that an attached medical unit like the 44th would only "need to know how to march on and off trains and ships." This assumption would lead to a near-catastrophe for the unit. Although they received excellent training, it could not prepare them for all that lay ahead in the challenging environment of the Pacific War.

During World War II, U.S. Surgeon General Norman T. Kirk oversaw a medical organization larger than in any other conflict. Across both European and Pacific theatres, over 535,000 medics, 57,000 nurses, 47,000 physicians, and 2000 veterinarians would serve. Nearly 700 overseas hospitals were responsible for initial care of the wounded. In the U.S., 78 military hospitals cared for nearly 600,000 patients during the War.

The chain of care began with combat medics, two of which were generally assigned to each company. They provided initial care and determined whether a wound required evacuation of the patient to a Battalion Aid Station. If additional treatment were required, the patient was evacuated to a divisional Clearing Station, where the first formal triage of patients occurred and which also served as small surgical hospitals for urgent cases. Definitive care took place at one of the overseas General Hospitals or at a military hospital stateside, in the "Zone of the Interior (ZI)."

A General Hospital, such as the 44th, was to be one of the last stops in the Army's prescribed chain of operations. In particular, an attached medical unit, with experienced civilian doctors like the 44th's, were intended to treat some of the most complex medical cases. They would be positioned well away from the front lines of battle. A General Hospital would either return a soldier back to his unit, or determine if the soldier should be sent back home. The complete prescribed chain of evacuation was as follows:

Organic Medical Units—These units, consisting of medics or corpsmen, were attached to combat units and followed them into battle. The rapid response of the medics on the battlefield, under hazardous conditions, was responsible for saving many lives. Medics administered first aid, applied tourniquets, and injected pain-killing morphine. Medics and their litter-bearers moved the wounded to a Battalion Aid Station, often under fire.

Battalion Aid Station—Positioned one mile from the front, physicians and medics adjusted splints and dressings, and administered plasma and morphine. Soldiers also reported to the aid station for treatment of minor illnesses or mild combat fatigue.

Collecting Station—Located two miles from the front, near the regiment command post. Provided further adjustment of splints and dressings, administration of plasma, and treatment of shock. Wounded were evaluated and, if needed, were sent to a Clearing Station for transport to fixed hospitals.

Clearing Station—Located four to ten miles from the front. They treated shock and minor wounds, and grouped patients in ambulance loads for transport to Field Hospitals.

Field Hospitals—Positioned within thirty miles of a Clearing Station. Ideally, the wounded arrived within one hour of injury. Surgery was performed for the most severe cases. In the Pacific War, the Mobile Hospital (see below) was used more effectively.

Mobile (Field) or Portable Surgical Hospitals—These hospitals were a key innovation in the Pacific War, due to the sometimes-vast distances and environmental challenges of moving seriously wounded that needed more advanced care, like surgery. They were assigned to a theatre of operations and could be packed and moved quickly to follow the battle lines. They were complete hospitals with nursing care, surgical and medical wards, X-ray, laboratory, and pharmacy. The Mobile Army Surgical Hospital (MASH) units of the Korean War, depicted in the movie and television series, evolved from these as the result of the experiences of World War II.

Evacuation Hospitals—Treated illnesses and less urgent surgical cases. Patients could be reconditioned here to return to the front.

Fixed Hospitals—These hospitals were set up a safe distance from the front, either in the theatre of operations or stateside, and they tended to remain in one location for longer periods of time. Typically, they were housed in prefabricated buildings or in churches, schools, and other public buildings that became part of the War. The three main fixed hospitals included:

Station Hospitals—Usually attached to a military base, they were designed to treat illnesses and injuries among personnel stationed at that base.

General Hospitals—The largest facilities where patients received long-term treatment. They sometimes grouped special treatment units in a large complex. Some of the General Hospitals were specialized for certain types of wounds or illnesses, such as for craniocerebral, spine, eye, chest, or neuropsychiatric care. They were staffed by more experienced physicians and supporting staff.

Convalescent Hospitals—Designed for rehabilitation of the severely wounded soldier who would receive a medical discharge. This type of hospital was also a World War II innovation.

In the Pacific Theatre of Operations, Surgeon General Kirk served over 1.5 million troops deployed across the vast area. There would be a total of 146 hospitals, including 15 Evacuation Hospitals, 24 Field Hospitals, 46 General Hospitals, and 61 Station Hospitals. The mortality rate for the wounded dropped significantly under Surgeon General Kirk's leadership. His experiences from World War I led to many advancements, including improvements in military trauma care, guidelines for amputation, and in techniques for limiting infection. Drug advancements, including the use of sulfa powder (sulfanilimade) for treating wounds, the discovery of penicillin, and atabrine for malaria prevention would also prove to be effective.

The efficiency and effectiveness of U.S. military medical operations was a key factor of success in the War. Even outbreaks of disease, as experienced in the Pacific War, could severely weaken a fighting force and potentially lead to defeat.

It's an ironic part of war, that the mission of the medical services is to return a soldier back to combat as soon as possible, and evacuate back home only if necessary. I suppose that this makes sense, both soldiers and doctors pledged to do their respective jobs in the conduct of war.

In February of 1943, Mom announced that she was expecting, the likely due date being in late August or early September. She returned to Kansas City and moved in with her parents while Dad prepared for duty overseas. Dad was promoted to 1st lieutenant on April 13, 1943. The couple welcomed the modest increase in pay as they prepared to have their first child.

Much like Camp Barkeley in Texas, Ft. Sill in the Oklahoma summer months was miserably hot, humid, and windy. With no air conditioning, many would soak their sheets in cold water before bed time, in order to gain some coolness from condensation. The long hikes with limited water were brutal. But, learning to cope with the heat and humidity would pay off in the future.

Founded in 1869 as Fort Wichita, the Oklahoma military post was later renamed Fort Sill. The fort had a colorful past, particularly during the Indian Wars of the late 1800s. It was established with the help of the legendary "Buffalo Soldiers" of the 10th Cavalry Regiment. They were an African American cavalry unit of ex-slaves formed after the Civil War by white Army Colonel Benjamin Grierson. Grierson, who selected the site for the fort, had members of the 10th Cavalry construct many of the stone buildings that still stand today. The Apache Chief Geronimo was held there after he surrendered. He shot his last buffalo while at the fort. Following World War I, the fort was also the site of experimental weapons testing, including the U.S.'s own version of a "balloon bomb." The test of one balloon bomb called "The Sausage" went terribly wrong. It exploded prematurely at lift-off, releasing its flames and toxic gas on troops standing below. A number of men were killed and the program was subsequently shut down.

On September 2, 1943, Lieutenant Odrowski received a telegram announcing that his wife had gone into labor. Mom liked to tell the story of the birth. Her father, a butcher by trade and owner of the corner grocery, exclaimed confidently that he would be responsible for taking her to the hospital when the time for delivery came. When Mom walked up to him saying, "Papa, it's time to go," she said that he turned white as a ghost and sat frozen and speechless in his chair. Her mother called younger brother Chester, who came over and took his sister to the hospital. Edward Michael Odrowski (Eddie) was born that evening.

Dad paced nervously around his desk at Ft. Sill, waiting for his phone to ring. His drill instructor, Sergeant Ducote, helped pass the time with a few hands of gin rummy. Sergeant Lawrence "Swift" Ducote was a high school teacher from Cottonport, Louisiana. He was Dad's top sergeant and close friend throughout the War. Ducote's wife Narcille and Mom were also good friends, as they spent time together in Lompoc and in Texas. Finally, the phone rang and his brother-in-law Chester gave him the good news. He was ecstatic at the birth of his son, Eddie. Dad and Ducote celebrated with some shots of good whiskey. Afterwards, the sergeant went into town and bought out the local cigar shops.

The next morning when the enlisted men lined up for morning announcements and drill, the tough first sergeant from Louisiana called "attention." Dad walked up to the platform where he and Ducote looked down upon the 500 or so men. Ducote continued in his southern drawl, "Listen up. Yesterday the lieutenant's wife gave birth to their son. This morning we're going to take some time to celebrate. The corporals will pass out cigars and matches. I better see every one of you smokin' that cigar. If I don't, it'll be yo'ass today!"

When Dad told this story, he always enjoyed imitating the sergeant's southern accent. He said that a lot of the EMs were just out of high school and many had never smoked a cigar before. Many of them turned green and ended up sick on the grounds. The local Lawton, OK, newspaper published an article that described the celebration, stating "a large cloud of smoke could be seen rising above the troops at the fort."

In September of 1943, the 44th was notified of their deployment overseas. Due to security reasons, the location of their assignment was unknown to them. The unit was told that they would report to Ft. Stoneman, just up the Sacramento River from San Francisco. Stoneman was a major jumping off point for troops deploying to the Pacific Theatre.

Dad returned to Kansas City for a short leave with Mom and their infant son. The time passed too fast for the couple. At the time, the outcome of the War on both fronts was very uncertain. It was stressful not knowing when or if they would be reunited. There was much to discuss when facing the realities of a spouse going to war. Mom would continue to live with her parents while Dad was overseas. Dad would send money home. After emotional good-byes with his wife, son, mother, father, and sisters, Dad headed back to Ft. Sill. He was driven to Union Station in Kansas City by his older brother, Stanley, nicknamed "Bud."

Bud was 10 years older than Dad. Dad looked up to him as a mentor and friend. As they pulled up to the station, Bud parked by the curb. Usually upbeat and joking with his younger brother, he was unusually solemn. He reached under the front seat of the car and pulled up a fifth of rye whiskey. The brothers drank a farewell toast in the car. Bud said, "Now Slim, don't you do anything stupid over there. I want to see you back here in one piece. We've got a lot of good times ahead of us."

Bud, who had backed Dad in more than a few neighborhood fights, might have wished that he was going instead of his younger, ambitious brother. He watched him walk away and blend into the other troops as he entered the station. In spite of the concern, I'm sure that he also felt some pride. Bud had worked on the railroad since he was 17 years old. He was over the draft age and had been injured in an accident as a brakeman. Though not officially deferred from the draft, many officials rejected railroad workers, as they were considered essential on the home front in wartime. During the War, railroad workers would haul troops and materials in record volumes as war production ramped up. The capacity and efficiency of the railroads would prove to be a strategic advantage for the U.S.

The large train depot in the central part of Kansas City had been transformed into an Armed Services dispatching center. It buzzed with the energy of men and women from all branches of the service, in uniform and carrying duffel and sea bags. The dining and waiting areas were filled with service people. The USO set up a room to assist with travel issues and to provide free coffee, donuts, and soft drinks. For many young men and women just entering the service, it would be their first long trip from home. Many melancholy goodbyes could be observed in the lobby and alongside the departing train platforms.

At Ft. Sill, final preparations were made for deployment overseas. The logistics of moving a fully staffed medical unit was a challenge. Trucks were loaded and taken to the train depot in Lawton, OK. The 44th left Ft. Sill by train bound for Camp Stoneman, near Pittsburg, CA. Although they didn't know their ultimate destination, going to the San Francisco Bay area likely meant deployment to the Pacific. The long journey was prolonged by having to side track the westbound train to allow other trains to pass. If the delay was a long one, the MAC training officers would lead groups out for calisthenics or other drills. A layover at the train station in Salt Lake City, UT, provided the opportunity for Dad and the MAC officers to lead the doctors, nurses, and enlisted men on a march around the city. They completed a six-mile march in formation from the train depot to the Mormon Temple and back. Bystanders were impressed by the precision of this sharply dressed marching unit, cheering and wishing them well as they passed by.

The 44th arrived at Camp Stoneman on September 20th. This major staging site for troops heading to the Pacific War was located approximately 40 miles from San Francisco in an area known as the California Delta. The San Joaquin and Sacramento Rivers converge here and flow into the Bay Area. A famous paddlewheel steamboat, the *Delta Queen*, was converted into a troop carrier to ferry the men some 60 miles by water to the Embarcadero at San Francisco. Built in Scotland in 1924, the *Delta Queen* was one of the most lavish and expensive steamboats ever built. The members of the 44th enjoyed one last luxurious ride down to the San Francisco Bay. The *Delta Queen* was used extensively by the Navy during World War II. It was later taken to New Orleans, where it exists today, designated a National Historic Landmark.

The 44th was given instructions to gather their personal gear and report to the docks on San Francisco Bay. The ship they were assigned to was originally called *America*. It was built in 1938 and was intended to be a luxury liner for travel between the U.S. East Coast and Europe. As the War in Europe began, it was commissioned into the U.S. Navy and renamed the U.S.S. *West Point*. It would be used extensively to transport troops from California to Australia. It could carry 10,000 troops and had a crew of 2,000.

Hauling their duffel bags, the 44th, including officers, doctors, nurses, and support staff, boarded the *West Point* on September 24, 1943. The moment was described as melancholy, as individuals pondered the uncertainty of what lay ahead of them. They were now leaving home and their loved ones, not knowing when or under what circumstances they would come back. The members of the 44th lined the decks as the ship passed under the Golden Gate Bridge. This was the last view of home for troops departing for the Pacific War. In the foggy distance, the iconic bridge disappeared from view.

As the 44th went out to sea, they wondered what their ultimate destination would be. Secrecy was important to avoid leaking troop movement information to Japanese Intelligence. A Navy destroyer and a blimp accompanied them partially out to sea for protection from Japanese subs that patrolled down the West Coast. At a certain distance offshore, the escorts disappeared. The 44th's members were not sure if that should be comforting or disturbing. It indicated that they were in international waters and now part of the war zone.

The ship zig-zagged across the vast Pacific in order to evade Japanese subs and their torpedoes. The erratic route taken made for many more long days of boredom aboard ship. The excitement and adventure of being out at sea diminished after a few days. Many had never been this far from home. It's easy to imagine the feeling of homesickness that ensued. As the temperature got increasingly warmer after a week at sea it was obvious that Alaska was not their destination. A pilot on board deduced that the ship had bypassed Hawaii. Instead it went further south within Polynesia, moving between the islands of Samoa and Tonga.

Life aboard ship on the long journey was mostly uneventful. Activities included sunning on deck, playing cards, reading, writing letters, and talking. There were occasional life boat drills. A life jacket was a constant companion. The jackets were called "Mae Wests" by the troops since they gave everyone a shapely hourglass figure. There were many restrictions on board to decrease the risk of enemy attack. Portholes were blacked-out and smoking was prohibited on deck after dark. Trash could not be thrown overboard since it could indicate the presence of a ship to enemy spotters. The Japanese Navy with its ships, submarines, and carrier-launched planes still controlled the Pacific.

Accommodations were better for the officers than the enlisted men. Officers shared converted state rooms. Enlisted men were stacked high in bunks below deck, where it was cramped, hot, and smelly. Sleeping outside on the deck was preferable. Water was rationed for everyone and only turned on at certain times of the day. You had to shower quickly or risk having the water cut off. There were two assigned chow times per day per person. The tables in the mess area had ledges on all sides. In the event of rough seas, the ledge kept your plate and food from sliding off the table. Seasickness and boredom were common over the long overseas trip. Cramped quarters and a long time at sea took its toll. Tempers flared below and fights were common. Discipline and work details were necessary to maintain order and cleanliness. Even Army officers could be cited by the Navy for minor violations. They were typically punished by cleaning the decks or dishwashing duty in the galley.

The ship crossed the equator on October 1, 1943. Following a traditional sailor's ritual, every ship crossing the equator held a large celebration to initiate first-timers into the "Domain of Neptunus Rex." This provided some good fun to break up the monotonous journey. After crossing the equator, "Pollywogs" (Army initiates) received subpoenas to appear before King Neptune and his court (Navy personnel) who officiated at the ceremony. This is preceded by a beauty contest of men dressing up as women, each department of the ship being required to introduce one contestant in swim-suit drag. All were encouraged to dress up in some outrageous costume. With some "dark humor," two of the 44th's doctors, dressed in operating smocks blood-stained with ketchup. At each of their sides was strapped a large handsaw. Many onboard were "interrogated" by King Neptune and his entourage, with the use of "truth serum" (i.e., hot sauce & aftershave) and whole uncooked eggs put in the mouth. During the ceremony, the "Pollywogs" would undergo a number of increasingly embarrassing indignities (e.g., crawling on hands and knees on nonskid-coated decks, being swatted with short lengths of firehose, being locked in stocks and pelted with mushy fruit, getting dunked in a water coffin of salt-water and bright green sea dye, crawling through chutes or large tubs of rotting garbage, or kissing the Royal Baby's belly coated with axle grease). This was very entertaining for the experienced Navy "Shellbacks." At the end of the event, everyone received a certificate stating that they had been initiated into the "Solemn Mysteries of the Ancient Deep."

On the morning of October 10th, a large land mass was observed in the distance. The outline of a city took shape as the ship got closer. The ship entered a beautiful bay with brilliant blue water. Homes and buildings along the cliffs had red tile roofs. Some onboard recognized the arched Sydney Harbor Bridge ahead. After seventeen days at sea, he 44th had arrived in Sydney, New South Wales, Australia.

Australia

Australia was an exciting destination for the many U.S. troops that arrived there from the end of 1941 through 1943. The vast, diverse land had many natural wonders including tropical beaches, offshore reefs, deserts, and the areas known as "the bush" and "the Outback." It's a land of exotic animals, such as kangaroos, wallabies, koalas, crocodiles, and kookaburras. It was also the land of roasted mutton and warm beer. Dad's main pet peeve during World War II wasn't the Japanese snipers, kamikazes, rain, heat, mosquitoes, or tropical diseases. There was one topic that would always "set him off" and that was, *warm beer*. He complained that when they received their beer rations in Australia, it was always warm. During wartime, there were practical reasons for this. In the tropical South Pacific, ice was practically non-existent. Refrigeration was scarce and needed for essentials like preserving whole blood and medicine. Hence, warm beer. And to put it mildly, Dad was a "beer guy." As the 44th's time in the Australian summer went on, the thirst for an ice-cold beer got more intense.

It's actually a myth that warm beer was preferred by the Australians. Sure, some of the British stouts and ales were typically drunk at room temperatures. But lager beers came to Australia in the 1800s. As in Europe, they were processed and consumed at cooler temperatures, which was a good match for the hot Australian summers. They steadily grew in pop-ularity up to the time of the war.

The 44th operated a large hospital at Townsville, Australia, in the more tropical northeastern province of Queensland. The U.S. Fifth Army Air Corps operated the airfield next to the hospital. The Air Corps supported the battle for the Solomon Islands and New Guinea with fighter and bomber support, and the transport of supplies. On return flights, the transports would bring back wounded to the hospital. The 44th's doctors were supported by a full female nursing staff. The local Air Corps pilots and crew were eager to invite the nurses of the 44th to the movies and dances that they sponsored. But due to safety concerns, the MAC officers had set up tight security measures around the nurse's camp. Under 24x7 guard the nurse's safety and whereabouts were closely monitored. Also, naturally, the Army staff was protective, particularly as it related to the "fly boys" of the Air Corps.

What does this have to do with the problem of warm beer? It led MAC Mess Officer Walter Teague to come up with a brilliant solution. As he watched an Army Air Corps C-47 cargo plane take off, circle out over the ocean, and return on a test flight, an idea flashed in his mind. If beer could be flown up to 10,000 feet or more it would be cooled to a perfect drinking temperature. It seemed logical, the Air Corps had the planes and the 44th had the nurses. Could a deal be made to allow the airmen to call on the nurses? The plan was put into motion. When a maintenance test flight was scheduled, the 44th's Mess Officer would be called. Cases of beer were driven out to the air strip in an ambulance and loaded into the transport plane. Thirty minutes at 10,000 feet was all it took. There were disciplinary risks to the airmen for collaborating, but this didn't seem to be an issue. They were even willing to face the risk of being shot down by Japanese Zeroes as they allowed time to cool down the beer.

Dad also had a dislike for Australian roasted mutton. A sheep less than a year old is butchered to get lamb, an adolescent sheep's meat is called hogget, and an adult sheep or goat provides mutton. Mutton is tougher and has a stronger flavor than lamb. During the Great Depression Dad's family was forced to butcher their goat. Evidently, Dad had been pretty fond of the goat. He said that his mother boiled the old goat for hours. It smelled up the house for days and the meat was still tough as shoe leather. When offered roast mutton by Australian officers, Dad had to politely refuse. One smell and he said that his stomach turned.

But out of the places Dad went to during the War, Australia was his favorite. It was exotic and adventurous. He described the land as vast and wild, reminiscent of the American West he visited in the 1930s. He'd go horseback riding with some Australian Army officers, far enough into the bush to see kangaroos, their smaller Wallaby cousins, and koalas. Dad described the strange sight of seeing a group of kangaroos jumping across the beach and the grasslands. He also described watching the kangaroo males "box" each other. Although mostly docile towards humans, the kangaroo can be aggressive if provoked. They use their forepaws like a boxer, then kick their opponent with powerful hind legs and sharp feet. The Aussies would occasionally shoot a kangaroo for meat. Dad said that it was very good, the steaks were lean, like filet mignon. He much preferred it to the mutton.

Upon arriving at Sydney, the 44th was given some liberty time to go into the city and let off some steam in the local pubs. Staging and awaiting orders, the members of the 44th were able to explore the city. Some took an excursion to Bondi Beach, one of the finest near Sydney with white sand and clear, warm waters. Shark nets, dating back before the War, are positioned offshore to try to keep the man-eating species away from the beach.

Although the Japanese had attacked the Australian port of Darwin in an air raid similar in nature to the attack on Pearl Harbor, Australia was a fairly safe place to be. A full-scale invasion was originally planned by the Japanese. But as the Australian and U.S. Armies joined forces to fight the Japanese on New Guinea, Japanese invasion of the "land down under" was highly unlikely.

The 44th's stay in Australia was intended to be a short one. Their orders came, identifying that they were to go to New Guinea where the heaviest engagements with the Japanese were occurring and casualties were mounting. On October 23rd the men of the 44th boarded a train for Brisbane, a city 600 miles away to the north. At Brisbane, the 545 men of the 44th boarded a Liberty ship called the *Peter De Smet*. The nurses stayed behind to join up later at the hospital location in New Guinea. Liberty ships were cargo ships built in the U.S. to support the massive movement of troops and supplies during World War II. Another 2,500 men boarded six other Liberty ships, all bound for Milne Bay, New Guinea. Moving off the Gold Coast of Australia into the Coral Sea, the sights were incredible—azure blue waters, flying fish, and large sharks were seen offshore. There were risks to the journey. The Great Barrier Reef, off the Queensland coast, is made up of a number of individual reefs that pose a navigation issue for ships. Precautions were also made to protect the convoy from possible Japanese attack by sea and air. Naval destroyers escorted the ships and a number of PT (Patrol Torpedo) boats kept a lookout for Japanese submarines. Future President John F. Kennedy, a young Navy lieutenant, commanded the famous PT-109 in the area of the Coral and Solomon seas earlier in 1943. The skies were clear and the seas calm as the ships made their way north towards the Solomon Islands and New Guinea.

As night fell on December 18, 1943, some of the men of the 44th made their way to the top deck to take in the tropical breeze and gaze at the luminescent marine life in the ship's wake. Some smoked and played cards below deck under dim lighting. As many were getting into their bunks or walking through the narrow passageways, the ship was rocked as if hit from below. A deafening grinding sound was heard. The ships engines wound down. A warning signal sounded and an announcement was given, the men were ordered to stay below until further notice. Some thought that the ship may have been hit by a torpedo. Soon, fears were somewhat eased as word came around that the ship had hit a reef.

The Bougainville Reef sits in the northern Coral Sea due east of Cooktown, Australia. A navigational error had caused all seven ships to run aground. The ship's captains conferred on what to do. They contacted the U.S. Navy. It was determined that it would be a high risk to attempt a rescue at night. So, they would wait until dawn. The U.S. Navy contacted the Australian Navy who would assist in the rescue the next day. It was an uneasy night for the passengers of the ships. Like sitting ducks, they were stuck in the middle of the ocean on top of the reef. PT boats kept a night-long patrol around the ships.

The men onboard were relieved to see the early light of dawn. As the sky lightened, the massive size of the reef could be seen in the sea below. Stretched across a long section of the reef were all seven Liberty ships, including the 44th's *De Smet*. At dawn, five RAN Bathurst-class Corvettes, the *Castlemaine, Lithgow, Gympie, Stawell*, and *Gladstone* arrived to aid in the transfer of men and cargo. Small rescue craft were lowered from the stranded ships. As the rescue craft would align next to the ship, eight men at a time would climb down a net ladder. Swells of four to six feet that morning made for a risky manoeuvre. The men would time the motion of the swell, then jump into the small boat.

Dad recalled his experience, "A large swell came up, lifting the ship up. The weight of the men caused the rope netting to swing out. When it came crashing back, my knee slammed into the steel hull." Dad was affected by the knee injury throughout the rest of his life.

Once in the rescue craft, the men then boarded the PT boats. The PT boats then ferried them to the RAN ships. The RAN vessels then proceeded to a Naval base at Cairns, Australia, back on Queensland's Gold Coast. The 44th was back in Australia. As they found out later, the *De Smet* was also transporting over 800 barrels of high-octane aviation gasoline and ammunition in the bottom holds, very hazardous cargo. A Japanese torpedo or aerial bombing would have been devastating. As Walter Teague later commented, "One Japanese torpedo, and we would've been blown to the pearly gates." Fortunately, no Japanese sub patrols or fighter planes were in the area to take advantage of the vulnerable ships.

The 44th awaited new orders while at Cairns, Australia. They were told that they would now be sent to Black River, near the city of Townsville, North Queensland. There, they would set up a 1,000+ bed hospital. It would replace a former Marine Corps base, as the Marines had been deployed to fight in the islands to the north. Once the hospital was set up, the nurses would be relieved of temporary duty at Brisbane and join the men there.

The hospital at Black River had many large wooden buildings with corrugated tin roofs. It would operate as a fixed General Hospital with a capacity of 1,000 beds. Barracks were partitioned with two to a room, with separate shower and latrine areas. Many of the buildings were "open air," without screens, which led to issues with flies and other bugs. The hospital area was close to the beach. The golden and white sand beaches of northern Queensland are some of the most beautiful in the world. Crystal clear blue water abounds, with colorful tropical fish, as the massive Great

Barrier Reef lies offshore. Some of the 44th's staff took trips to Magnetic Island, a tropical paradise that was about 10 miles offshore. Fish and wildlife also abounded in the nearby Black River which flowed into the Coral Sea. Crocodiles had to be respected as they lurked in the brackish river.

The 44th had many of the comforts of home at the hospital. Major Maurice Richter, MD, a former college athlete, oversaw the construction of baseball, volleyball, and tennis courts. In baseball, the enlisted men usually beat the officers. In volleyball, the officers beat the enlisted men. Dad had been nursing a knee injury he sustained during the rescue when shipwrecked, but was back in action on the baseball field and volleyball court.

Horseback riding into the "bush" and along the beaches was also a great diversion. Some of the 44th's staff from Texas even joined some Aussie "cowboys" in an Australian version of a rodeo.

The men and women of the 44th were made up of all religious faiths. Worship was important to them. The men of the 44th constructed a chapel for worship services. The 44th had two chaplains who presided over Sunday services, Lieutenant, Father James F. Hayes, Catholic priest, and Lieutenant, Reverend Elijah G. Willis, Protestant minister. A number of Jewish men and women were also part of the 44th, but they had no official rabbi present.

At Townsville, the 44th created an "open air" movie theatre. First-run movies were shown to the staff and their patients, many seen before being released to theaters in the States. The most popular movies of 1943 included *For Whom the Bell Tolls*, *Casablanca*, *The Song of Bernadette*, and *This is the Army*. The top stars of the day included Gary Cooper, Humphrey Bogart, Bob Hope, Betty Grable, Bing Crosby, James Cagney, Ingrid Bergman, and Clark Gable. At the dances held at the Officer's Club, music from the Big Bands dominated. Glenn Miller, Harry James, Benny Goodman, and Tommy Dorsey, with Frank Sinatra's vocals, provided the soundtrack.

Dad spoke highly of the Aussies. He admired their independent spirit and courage. He said that they fought tenaciously and were "good blokes" to have on your side in a fight. But there would be mounting tensions as large numbers of American troops arrived in Australia. One reason for the resentment was the fact that many Australian men had been away fighting the Germans in North Africa. Many were being called back for duty in New Guinea in an effort to respond to a possible Japanese invasion of Australia.

The Japanese stoked the fears and jealously of the Australian men fighting in New Guinea. They dropped pamphlets from planes over troop positions including graphic cartoons of American GIs taking liberties with the women at the home front. To the Australians the Americans were "overpaid, oversexed, and over here." This led to confrontations between Aussies and Americans in the bars frequented by military personnel. MPs patrolled to try to keep the peace, but sometimes they would also get involved in the fights. To some of the American soldiers the absence of Australian men presented an opportunity. One 44th veteran noted, "As a rule, the Sheilas (nickname for Australian women) preferred the enlisted men to officers." Another veteran commented, "The Australian women were pretty, but many had bad teeth." Maybe he was a dentist?

U.S. Armed Forces were strictly segregated at the start of World War II. Although some combat units were formed, most black troops were relegated to support tasks, such as transporting supplies as truck drivers, road construction, kitchen duty, sanitation, and other supporting services. During the War, more than 2.5 million African American men registered for the draft. African American women also volunteered in large numbers. When combined with Black women enlisted into Women's Army Corps, more than one million African Americans served in the Army during the War. At home, the shortage of labor was a critical threat to the war effort. African Americans filled much of the gap along with American women, who fought for their right to work in jobs they were previously denied access to.

Racial tensions flared up in Australia as Black troops arrived by the thousands. Discrimination imposed by white officers spurred incidents with Black troops. The most violent occurred at Townsville on May 22, 1942, previous to the 44th's arrival there. Referred to as "The Townsville Mutiny," it involved African American troops from the 96th Battalion, U.S. Army Corps of Engineers, who were stationed at a base outside of Townsville, called Kelso Field. They were a construction battalion that built bridges and barracks. The troops had been the subject of regular racial abuse by some of their white officers. It was believed that a black sergeant had died at the hands of a white superior. Outraged troops of Companies A and C sought to kill the white officer responsible for the sergeant's death. The black troopers began firing machine guns at the tents of white officers. The resulting siege lasted for a full day. There was at least one fatality and many wounded. Australian soldiers were called in to help stop the riot. An American journalist wrote a report on the mutiny but it was suppressed by the military. In an interesting footnote, future U.S. President Lyndon B. Johnson, then a young congressman, was on an official tour, visiting Townsville at the time.

Overall, African American troops have commented that, for the most part, they were treated cordially by the Australians. Many stated that they felt more acceptance in Australia than in the U.S. During the Battle of the Bulge in Europe, General Eisenhower reluctantly formed an all-black volunteer infantry unit. They performed so well that Ike also used them in other key engagements. Other African American units served with legendary distinction, including the Tuskegee Airmen, the Buffalo Soldiers of the 92nd Infantry Division, and the Red Ball Express who drove supplies from the Normandy landing beaches inland under enemy attack. In 1948, President Truman signed Executive Order 9981, ordering full integration for the Armed Forces. By the end of the Korean War, almost all the military was integrated.

The safety of the Army nurses was a top priority. Early in the deployment to the Pacific theatre, a number of nurses become the victims of sexual assault. General MacArthur took swift action to put an end to the threat. He set a policy of strict security to protect their safety. The nurses were housed in separate compounds manned by heavily-armed security forces twenty-four hours per day. Their movements were to be protected by armed guards. Violence against the nurses would not be tolerated, carrying the full penalty of death for military personnel, if convicted. MacArthur knew that if the safety of the nursing corps was compromised, their presence in the theatre would not be sustainable. The nurses were vital to the treatment and recovery of wounded and sick soldiers.

While in Australia the 44th got firsthand experience with the treatment of battle casualties in the Pacific War. It was the goal of the Medical Corps to treat a wounded soldier within minutes of their injury. The sooner they could get to a soldier the better to relieve pain and avoid shock from blood loss. In the Pacific, medical treatment occurred across a harsh environment, from crushed coral beaches, to steamy mountain jungles, to muddy swamps. Wounded were treated in the field under fire, in tent hospitals a few miles from the front, and in more sterile hospital buildings hundreds or thousands of miles away. The complex chain of evacuation moved patients to where they could best be treated. At all points along this chain, decisions were made regarding when to treat, when to return to duty, and when to evacuate further to the rear.

As a military hospital, the 44th was expected to have protection from attack. The Geneva Conventions were a set of international agreements that originated in 1864 that established laws for the humane treatment of wounded or captured military personnel, medical personnel, and non-military civilians during armed conflicts. Article 19 states that medical units, i.e., military hospitals and mobile medical facilities, "may in no circumstances be attacked."

The Geneva Conventions came about from the experiences of the Battle of Solferino. On June 24, 1859, the armies of France and Sardinia fought Austrian forces near the northern Italian village of Solferino. The French were led by their ruler, Napoleon III, Sardinia by their ruler Victor Emmanuel II, and the opposing Austrians by their Emperor Franz Joseph II. It was the last major battle of Europe to be led by actual monarchs. This decisive battle in the struggle for Italian unity turned into a horrific bloodbath. In a single day of fighting, over 6,000 soldiers were killed and over 30,000 wounded. The wounded were shot, bayoneted, and bludgeoned. Men bled to death, starved, and died of thirst in the streets. The unsanitary conditions led to outbreaks of disease. Hospitals also came under direct attack by bullets and cannon.

Henry Dunant, a young banker from Geneva, Switzerland, arrived in a nearby town where wounded were being evacuated. The wounded occupied buildings, homes, and were laid out in the streets. There was not adequate water, food, medical supplies, doctors, or nurses to care for them. Many died waiting for assistance. Although not trained in medicine, Dunant abandoned his business activities and instead assisted the wounded. He distributed water, cleaned wounds, changed dressings, and even paid for supplies to be brought in from a nearby town. He enlisted the aid of charitable local women to assist the injured and dying. He wrote to his friends at home to ask for donations of supplies. He provided an example of how help could be organized so as to alleviate, as far as possible, the suffering which surrounded him.

Upon his return to Geneva, Dunant, still haunted by what he had experienced, wrote a book describing what he saw, called *A Memory of Solferino*. He published it with his own funds, and on the cover included the words, "Not to be Sold." Translated into multiple languages, it reached a wide audience. It horrified its European readers and made an immediate impact. More than just describing the inhumanity he experienced in the treatment of wounded, he ended the book by asking two questions, which appealed to the conscience:

> *Would it not be possible, in time of peace and quiet, to form relief societies for the purpose of having care given to the wounded in wartime by zealous, devoted and thoroughly qualified volunteers?*

This basic question became the inspiration for the founding of the Red Cross. But for volunteers to be able to carry out their relief activities during a war, they also had to be recognized and protected from harm. This led to Dunant's second appeal:

> *On certain special occasions, as, for example, when princes of the military art belonging to different nationalities meet at Cologne or Châlons, would it not be desirable that they should take advantage of this sort of congress to formulate some international principle,*

sanctioned by a Convention inviolate in character, which, once agreed upon and ratified, might constitute the basis for societies for the relief of the wounded in the different European countries?

This question was to result in the adoption of the original Geneva Convention. To this day, the Geneva Convention defends the principle of Medical Neutrality. According to the Physicians for Human Rights, this principle requires:

- The protection of medical personnel, patients, facilities, and transport from attack or interference
- Unhindered access to medical care and treatment
- The humane treatment of all civilians, and,
- Non-discriminatory treatment of the sick and injured

In principle, military hospitals were protected by the Geneva Convention. In practice, the Japanese government during World War II, never ratified the Convention. From their actions during their early conquest of the Pacific, it seemed that they did not plan to follow the intent of the Conventions either. Even in the war zones of the 21st Century, violations of this principle are still common. Hospitals and those treating the wounded are still known to be deliberately attacked.

The 44th prepared for their move to New Guinea. Memories of their first aborted voyage to New Guinea was still fresh in the minds of the 44th. The threat of Japanese submarines, dive bombers, and Zero fighters was still high. All personnel on ship were trained in "battle stations" and firefighting in the event of a Japanese attack at sea. Days onboard ship also included cleaning and kitchen duties. Anyone caught in even a petty offense, like littering, could be punished by kitchen duty or clean-up detail. It kept the men busy and probably helped ease the stress. Once again, as with all voyages across the distances of the Pacific, people killed the time at sea playing cards, reading, and writing letters home.

New Guinea

Mom was busy doing laundry in the basement. I knew that I had to act fast. Moving stealthily into the dining room I crawled up to the large bureau that rested against the wall. On my belly I moved under cover of the dining room table to the cabinet on the far right. As I quietly opened the door I was greeted by the smell of money. Lying before me were various bill denominations; cash stashed away for summer vacation. There were also stacks of savings bonds and bundles of canceled checks. But I wasn't after that kind of treasure. My target was a black, bound photo album. It contained Dad's photos from the War. He had shown me some of the less graphic ones before. Others, according to Mom, were off-limits to a 12-year-old boy. I opened the album to the back pages. There they were, graphic as hell. Dead Japanese soldiers, lying in mud somewhere. Some without heads. Some whose hands and feet seemed to be shredded off. I hadn't heard any stories to explain these scenes or why he had taken these photos. Then I noticed an object lying beside one of the dead soldiers. It looked like a Japanese sword that Dad kept in his closet. It was a prized memento of the War that he sometimes brought out for me to see. Some other photos caught my eye. They were of barely-clad natives in front of thatched huts on a beach. Dad and a few other officers stood behind them. The natives were short and dark-skinned with fuzzy hair. Men and women didn't wear shirts, just loin cloths. The men were thin and muscular. They smiled and looked relaxed. A naked native child stood holding his mother's leg. It was a strange scene, a contrast in cultures. It looked like Dad and the other officers had stepped back in time. I heard Mom walking up the creaky staircase from the basement. I quietly put the album back and retreated to my room.

Summer in Kansas City was mostly hot and humid. Dad worked eight hours a day in an air-conditioned office downtown. Going into work early, he always got home at 3:30 p.m. On summer days, he had his usual routine. He'd pop a beer and lay out in the sun on the concrete of the patio, sometimes so hot that you could fry an egg on it.

Another beer and soon he was napping in the summer sun. He called it "sunbathing." I think that it was stress relief. Since the War, Dad could sleep anywhere, under any condition. He could sleep through any noise, lying down, seated, even standing up. He said "That was the way it was in the Army. You slept whenever or wherever you could." Evidently, during the War, that included while artillery was firing or bombs were dropping. Refreshed after a short nap, he'd be ready to work in the yard, practice baseball with me, or help Mom with dinner.

That evening on the patio, after the sun set, I thought about the pictures I had seen that afternoon. Not wanting to incriminate myself, I didn't ask Dad about the dead Japanese soldiers. Not yet anyway. Instead I asked, "What were the natives like on the islands you went to?"

While on New Guinea, Dad described a trip he took with Colonel Weston, a few of the 44th's doctors, and his security detail. The men loaded into two jeeps and proceeded from their base near Hollandia, Dutch New Guinea (present day Jayapura), to a native village up the coast and on the other side of the mountains. As they left the beach area, they viewed areas that had been hit by heavy shelling prior to the U.S. landing. In a number of the coconut groves they passed, the tops of the coconut palms were gone. The tall coconut logs stood like sentries amidst areas of thick jungle growth. Two of Dad's sergeants armed with submachine guns scanned the road ahead. Four enlisted men armed with M-1 carbines rode along. The officers carried their .45 side arms. A native police officer accompanied them to help with navigation. He also knew enough Australian English to translate.

Although the Japanese had been mostly driven from the area, there were still some holding out in the jungle. Many were cut off from their divisions and near starvation. Those foraging for food were not likely to engage in an ambush. The jungle road was narrow, rough, and muddy. Dad and the others passed some native peoples walking along the road. Most smiled broadly at the men in the jeeps. Dad said that the natives would trade shells and various crafts for cigarettes and Army clothing. Some had dyed the Army khaki shirts brilliant turquoise and red. The translator told the Americans that the Australians made many rules to discourage contact and trading with the natives.

The jeeps made their way past a large lake (Lake Sentani). Some of the natives lived in huts with tall stilts that hoisted them over the water. Simple boats were used for fishing and navigating around the lake. A large flock of white Cockatiels flew from a tree by the edge of the water. As the jeep climbed into the mountains, various tropical birds could be seen and heard. Wild orchids were seen growing in the upper levels of the trees. The jungle was humid and had the musty smell of rotting vegetation. The jeeps passed fast-moving clear streams coming out of the mountains. Waterfalls could be seen below the road, flowing water cascading down large boulders and collecting in large, clear pools below. The tops of the mountains were cloaked in white clouds. It rained frequently in the upper elevations.

As the jeeps made their descent to the coast, a brilliant view of the ocean lay ahead. A strip of golden sand lay below in contrast to the clear azure water. Dad said that it looked like a tropical paradise. It was hard to believe that this peaceful scene was an area that had just been the site of brutal jungle warfare. It seems that nature is resilient if given a chance to recover from mankind's destructive ways. The jeeps came to an area where they had to cross a river to get to the beach area. The guide cautioned the men about getting into the water. Leeches were common and crocodiles were known to patrol the deeper areas. The crossing was made without incident except for one jeep getting bogged down in mud along the bank.

The jeeps were parked and the men walked onto the golden beach. A number of thatched huts were seen ahead. Dogs barked as they approached. Some children, naked and playing by the water, ran back to the huts. An elderly-looking man greeted them. The guide provided a brief introduction. The elderly man looked very serene and friendly. He motioned to follow as they walked toward a large hut in the center of the encampment. The officers asked if it was OK to provide some cigarettes to the adults and candy to the children. The natives welcomed the gifts. As the men approached they witnessed men, women, and even young children smoking American cigarettes.

In return for their gifts, the Americans were offered betel nut (a seed from the areca palm trees). Dad said the people carried gourds or coconut shells full of the crushed nut, harvested from the native tree. They'd wrap it in the leaf of the betel tree and chew it. It would provide a good buzz, much like strong chewing tobacco. It also turned the teeth of the natives dark red. Dad said that it had a peppery and bitter taste, not one he was fond of. Consumption of betel nut has been proven to have many harmful health effects, including being carcinogenic.

Dad described the natives, who he referred to as Kikis, as living a simple and slow-paced life along the beach. The older men and women were quite jovial and smiled and laughed a lot. Men only wore loin cloths and no shoes. The women did not wear shirts and young mothers openly nursed their children, some for a relatively long time, up to the age of three. Although their clothing was sparse, they also appeared to display modesty in other regards. The native community along the beach appeared to be very close knit. It seemed that men and women alike held positions of respect.

The doctors observed various skin diseases among the people. The guide told them that they are occasionally visited by some locally-trained native "doctors" who sometimes obtained medicines from the Dutch missionaries and Australians. It was exciting to come face-to-face with such a primitive culture. The natives of New Guinea who experienced the War in their remote corner of the world would never be the same.

The New Guineans had proven to be good allies. They had intimate knowledge of the mountainous terrain and were adept at living off the land. They quickly developed a dislike for the Japanese who invaded the island early in 1942. They perceived the Americans and Australians as friendly. Called the "fuzzy-wuzzy angels" by the Australians, many natives would serve as litter-bearers for moving the wounded down treacherous mountain pathways. Native men also engaged in tracking and harassing the Japanese. Calling upon their warrior traditions, they used their knowledge of the jungle and stealth to kill unsuspecting Japanese, sometimes using their traditional knives, arrows, and spears.

New Guinea provided a unique experience for the members of the

44th. With unspoiled mountains, beaches, rivers, reefs, and vast jungles, it was the among the most primitive and undeveloped parts of the world. New Guinea is the second-largest island in the world, next to Greenland. Located in Melanesia in the southwestern Pacific Ocean, it is separated by the shallow and 90-mile wide Torres Strait from its neighbor to the south, Australia. A large number of smaller islands surround it to the east, in the Coral Sea, and to the west, in the Arafura Sea.

The shape of New Guinea is often compared to that of a bird-of-paradise (indigenous to the island), and this results in the usual names for the two extremes of the island: the Bird's Head Peninsula in the northwest (Vogelkop in Dutch), and the Bird's Tail Peninsula in the southeast (also known as the Papuan Peninsula). A spine of east-west mountains, the New Guinea Highlands, dominates the geography of New Guinea, stretching over 1,000 miles from the "head" to the "tail" of the island. Many mountains have elevations of over 12,100 feet, making the island an impressive sight from offshore. Being close to the equator, most areas are hot and humid throughout the year, with some seasonal variation associated with the northeast monsoon season. Higher elevations are the exception, experiencing much lower temperatures. Vast southern and northern lowlands stretch for hundreds of miles. These possess lowland rainforests, extensive wetlands, savanna grasslands, and some of the largest expanses of mangrove forest in the world.

New Guinea is very biodiverse, with over 200,000 species of insects, between 11,000 and 20,000 plant species, and over 650 resident bird species. Most of these species are shared, at least in their origin, with the continent of Australia, which was until fairly recent geological times part of the same landmass

Netherlands New Guinea, the western half of the island, and the Australian territories, the eastern half of the island, were invaded in 1942 by the Japanese. The Australian territories were put under military administration and were known simply as New Guinea. The highlands, northern and eastern parts of the island became key battlefields. Papuan natives gave vital assistance to the Allies, fighting alongside Australian troops, and carrying equipment and injured men across the primitive land.

Approximately 216,000 Japanese, Australian and U.S. soldiers, sailors and airmen died during the New Guinea Campaign. Often overlooked today, the New Guinea campaign was the longest of the Pacific War, with 340,000 Americans fighting more than 500,000 Japanese. Soldiers in the Imperial Army of Japan had a saying: "Heaven is Java, hell is Burma, but no one returns alive from New Guinea." The jungle took a high toll on troops. Entire patrols would sometimes disappear in the New Guinea jungle, their skeletons found years later. For nearly four years, the Japanese struggled to hold onto the mountainous, jungle-choked island, fighting first against the Australians, then against Americans commanded by General MacArthur. Because holding onto New Guinea was central to the Japanese strategy for the war, they poured a vast number of troops, ships, and warplanes into the effort. Since they had to divert resources from other fronts, the battle for New Guinea contributed to Allied successes elsewhere.

New Guinea's jungles were thick and full of potential dangers, including pythons, large spiders, rats, and innumerable insects. The rainy season gave the air the smell of a rotting leaves or a musty basement. Mud caked everyone's boots. Mudslides were common, anything left outside could get swept away in a torrent of water and mud. Japanese were still in the area, but most had been cut off from their units and were near starvation. Some would attempt to raid U.S. food stores. Some would just wander into camp, unarmed, and turn themselves in for food. It was a difficult situation for U.S. troops; Japanese soldiers were known to act like they were surrendering, but could be holding a grenade to carry out a last-ditch suicide attack.

Dad had some photos of older men in khakis with long beards. He identified these as Dutch missionaries. They had been doing evangelical work to convert the natives to Christianity. When the War broke out and the Japanese invaded, many of the missionaries hid in the jungle with the natives. Some groups were captured and killed by the Japanese. Some were protected by the natives who knew the best places to hide and avoid capture. When the U.S. arrived some encountered patrols and were

39

escorted down to the Army medical units for treatment. Barely clothed, starving, and suffering from various diseases, they were cared for.

I can imagine that the Dutch missionaries read the Bible to the natives and preached the message of the Gospel. As the natives listened to the Book of Revelation and heard of the great battle of Armageddon, the description of the end times must have placed some memorable images into their minds. To the natives, the sight of the large invading Allied forces may have seemed like a prophecy come true. The great battle of World War II may have been viewed as the last battle of good over evil. Maybe they saw the invading Americans as the "good" force. Regardless, the technology, supplies, and military power of the U.S. must have been impressive. This led to the phenomenon of the "cargo cult." A "cargo cult" is a belief system among members of a relatively undeveloped society in which adherents practice superstitious rituals hoping to bring them modern goods supplied by a more technologically advanced society.

The vast amounts of military equipment and supplies that both Allies and Japanese airdropped (or airlifted to airstrips) to troops on these islands brought drastic changes to the lifestyle of the islanders, many of whom had never seen outsiders before. Manufactured clothing, medicine, canned food, tents, weapons and other goods arrived in vast quantities for the soldiers, who often shared some of it with the islanders who were their guides and hosts. This was true of the Japanese Army as well, at least initially before relations deteriorated in most regions.

The "cargo cult" exists in parts of New Guinea to this day. Followers believe that various ritualistic acts such as marching with bamboo "rifles," dressing in uniform, or building an airplane runway will result in the appearance of material wealth, particularly highly desirable western goods (i.e., "cargo"), arriving via airplanes.

The U.S. 6th Army invaded Hollandia, New Guinea, on April 22, 1944. The Allied forces, U.S. and Australia, were recovering from losses in previous battles. As a result, the Allies were not able to send in as many troops as they would have liked. The landings at Hollandia were undertaken simultaneously with amphibious landings at Aitape to the east. The battle was a success for the Allied forces, resulting in a with-drawal by the Japanese to a new strategic defensive line in western New Guinea and the abandonment of all positions in the east part of the island.

Casualties were on the rise and the hospitals were over capacity. New Guinea provided a good indication of what the Pacific War would be like. It was a harsh environment—tropical heat and humidity, rugged mountains and thick jungle, and incessant rains. The volcanic islands of the South Pacific had fresh water that collected in areas that bred mosquitoes. With the mosquitoes, came diseases, malaria being one of the most common. The jungle threatened to take an even larger toll than the Japanese bullets, bombs, and bayonets.

Malaria is a mosquito-borne infectious disease that affects humans and other animals. Malaria causes symptoms that typically include fever, tiredness, vomiting, and headaches. In severe cases it can cause yellow skin, seizures, coma, or death. A large percentage of U.S. troops in the Pacific would experience a bout with malaria. The drug Atabrine would largely replace quinine as a preventative treatment. The use of mosquito nets, topical repellents, and environmental treatments would help control the spread of the disease. The dangerous chemical DDT was used to kill mosquito larvae in standing water.

Among human parasitic diseases, schistosomiasis ranks second behind malaria in terms of impact in tropical and subtropical areas. Schistosomiasis, also known as snail fever and bilharzia, is a disease caused by parasitic flatworms called schistosomes. The urinary tract or the intestines may be infected. Symptoms include abdominal pain, diarrhea, bloody stool, or blood in the urine. Those who have been infected for a long time may experience liver damage, kidney failure, infertility, or bladder cancer. In children, it may cause poor growth and learning difficulty.

Dengue fever was another mosquito-borne viral disease occurring in the Pacific. Known to the troops as "break-bone fever," symptoms include high fever, rash, and muscle and joint pain, hence the nickname. In severe cases there is serious bleeding and shock, which can be life threatening. Those who become infected with the virus a second time are at a significantly greater risk of developing severe side effects.

Scrub typhus, also known as bush typhus, is a disease caused by a bacterium called "tsutsugamushi." Scrub typhus is spread to people through bites of infected chiggers (larval mites). The most common symptoms of scrub typhus include fever, headache, body aches, and body rash. The War showed that the disease was prevalent in damp areas covered with the very tall tropical kunai grass. Environmental treatments included burning or oiling areas of tall grass that harbored the mites.

Jungle rot was a particularly debilitating malady for the troops in the Pacific. Common in tropical climates, it is a chronic ulcerative skin lesion thought to be caused by polymicrobial infection with a variety of microorganisms, including mycobacteria. Ulcers occur on any exposed parts of the body, primarily on the back and legs. It may erode muscles and tendons, and sometimes, the bones. These lesions may frequently develop on pre-existing abrasions or sores sometimes beginning from a mere scratch.

Dad described the annoying "blow flies" on New Guinea. He said that you had to eat with one hand while waving your hand over your food with the other. If a fly landed on your food, it would leave a deposit that had nasty consequences, including diarrhea and high fever. Nurse Eda Teague, in a letter home, described the experience of New Guinea, "this place has the most vicious bugs I have ever seen. Some little ones about 100th of an inch bite like a rat. We have lots of rats here too. They are quite brave and they run through the tents all the time."

Besides these tropical diseases, most of them insect-borne, other common health issues would appear due to the conditions of living in the field. Impurities in food and water caused outbreaks of dysentery and diarrhea. Crude sanitation and decomposing bodies led to swarms of flies that carried impurities toxic to humans. The tropical heat and humidity would cause dehydration and heat stroke. Constant rains, wet clothing, physical exhaustion, and lack of sleep while lying in wet, muddy foxholes, would likewise take a physical and emotional toll. Not to mention the threat of Japanese attack, particularly at night, and noise from planes and artillery. Battle stress in the Pacific would be particularly severe and psychiatric care increasingly important. Emotional and physical stress would also take its toll on the doctors, nurses, and corpsmen of the military medical units. They witnessed horrific injuries, disease, and death on a daily basis. Many had mental breakdowns, during and after the War. Like health workers of today that operate in a crisis, they worked long hours without adequate rest. There was simply no one else around to do their jobs.

The Southwest Pacific Theatre of Operations provided a number of unique challenges, many not encountered in the War in Europe. In the European theatre, Germans retreated and surrendered in great numbers under the onslaught following D-Day. Not so with the Japanese, who felt divinely ordained to die rather than surrender. The numerous South Pacific islands lacked the amenities of the European cities and countryside. There were few large cities that offered any of the comforts of home. There was no cultivated land across most of the countryside. The environment would prove to be as dangerous as the enemy. Plus, the vastness of the Southwest Pacific theatre made the shipment of medical supplies and the evacuation of wounded a real challenge.

The Army Medical Corps had to learn quickly how to tackle the unique conditions and maladies of the Pacific. General MacArthur realized the importance of disease prevention and treatment. Disease threatened to disable his troops faster than the Japanese could. He formed a number of task forces and called in many experts from around the world. Improvements were made, but over time the number of disease-related casualties would still eclipse those related to battle.

To ease the tensions of the medical staff, there was some fun competition between 54th and 44th hospitals in volleyball and baseball. One veteran said that when a baseball was hit into the jungle, it was given up for lost. Once, the men were surprised when a ball got mysteriously thrown back onto the field. No one questioned who was out there. It may have been Japanese soldiers taking a break to watch some baseball—a memory of better times at home.

There had been ongoing debates regarding what the next major invasion target would be following New Guinea. Admiral Nimitz of the U.S. Navy pitched for Formosa (present-day Taiwan). General MacArthur stuck bullishly to his commitment to return to the Philippines. At a meeting with President Franklin Roosevelt in Hawaii, each made his case. Scouting reports from the Navy's Admiral Halsey indicated that the island of Leyte was lightly defended. MacArthur's argument won out with FDR, and the decision was made to move forward with the invasion of Leyte and the Philippines. With Leyte in hand, the U.S. could cut the Philippines in two and have a strong position to support the invasion of Luzon, to the north, and the prized city of Manila.

New Guinea had been a surreal experience for the men and women of the 44th. They had been transported to another world—the primitive Kikis chewing their beetle-nut, naked, living in grass huts on the beach. The jungle with torrential rains and mud. Tropical birds, giant bats, jungle rats, bugs, snakes and giant crabs. Plus the threat of the enemy snipers.

The officers of the 44th planned their next objective, support for the Leyte Invasion. The 44th was expected to land on November 18, A-Day (first landing) plus 27 days. By that time, it was expected that the U.S. 6th Army would have cleared the Japanese from Leyte's east coast and Central Valley. The 44th would set up their hospital at an inland location to support the troops as they drove the Japanese to the north. The nurses would report for duty at the 54th Station Hospital in New Guinea to wait for deployment to Leyte. They would join the hospital later, once the 44th's inland position was secured.

The Philippine Islands

Located in the Pacific Ocean near the equator, the Republic of the Philippines consists of over 7,500 islands that form an archipelago. Approximately 2,000 of the islands are inhabited. The country can be divided into three main areas: Luzon (the largest, northernmost island, which includes the city of Manila); a group of islands called the Visayas (including the major islands Panay, Negros, Cebu, Bohol, Samar, Masbate, and Leyte); and Mindanao, the second-largest island in the Philippines, found at the southern end of the archipelago.

The Philippine archipelago is bounded by the Philippine Sea to the east, the Celebes Sea to the south, the Sulu Sea to the southwest, and the South

China Sea to the west and north. The Philippines have a total coastline of over 22,000 miles. The islands spread out in the shape of a triangle, with those south of Palawan, the Sulu Archipelago, and the island of Mindanao outlining (from west to east, respectively) its southern base and the Batan Islands to the north of Luzon forming its northernmost point.

The Philippines abound with diverse, natural beauty. There are mountain ranges, coastal plains, large river systems, volcanos, hot springs, and spectacular lakes. The islands are composed primarily of volcanic rock and coral, but all principal rock formations can be found. The mountain ranges typically run north to south, like a spine that divides an island into eastern and western sides.

The ethnically diverse people of the Philippines collectively are called Filipinos. The ancestors of the vast majority of the population were of Malay descent and came from the Southeast Asian mainland as well as from what is now Indonesia. Contemporary Filipino society consists of nearly 100 culturally and linguistically distinct ethnic groups. Of these, the largest are the Tagalog of Luzon and the Cebuano of the Visayan Islands. Filipinos are predominately Roman Catholic. Various tribal groups still exist in remote areas across the expanse of the Philippine Islands. Some of these groups maintain their traditional religious beliefs. Prior to the War, headhunting and the ritual sacrificing of pigs and chickens was still practiced.

Spain colonized the Philippines in the 1500s. The Spanish had a lasting influence on the Filipino culture, including religious life (Catholicism), language, food and architecture. The Spanish-American War in 1898 ended Spanish rule of the Philippines. The Spanish were obligated to sell the Philippines to the United States for $20 million. The Philippines became a U.S. territory with the signing of the 1898 Treaty of Paris and the defeat of the Filipino forces fighting for independence during the 1899–1902 Philippine-American War.

Nipa huts were the native houses of the indigenous people of the Philippines before the Spaniards arrived. They were designed to endure the climate and environment of the islands, made from local plant materials like bamboo, with roofs covered in palm thatch. The accessibility of the

45

materials made it easier to rebuild if damaged by a storm or earthquake. In the rural areas, houses are often small and simple, consisting of just one or two rooms, and are elevated on piles. The piles lift the living areas above areas that flood. The open spaces below the structures are used to store tools and other household belongings, as well as live chickens and other smaller farm animals. Especially in the fishing communities of coastal regions, houses are typically raised above the ocean, river, or floodplain to accommodate boat traffic and the ebb and flow of the tides. There are often elevated networks of walkways linking homes within a community.

At the time of the War, the Philippine economy was highly agricultural. The hilly and mountainous regions of the islands have moist, fertile soils, often with a significant concentration of volcanic ash. The rich soils support the growth of tropical fruit trees and pineapples. Oil palms, vegetables, and other crops are grown in the peat-like areas, as well as in the younger, sand-based soils of the coastal plains, marshes, and lake regions. Dark, mineral-rich fields over much of the Visayas, and the northwest tip of Luzon grow coffee and bananas. Rice is a staple in the Filipino diet. About one-fourth of the total farmland is utilized for rice growing. Terraced hillsides, some from ancient times, mark the landscape. At the time of the War, Filipino women pounded rice in front of their homes, as they had for centuries. Other areas in the central and southern Philippines grow cassava (manioc) and sugarcane. Dense forests provide for timber harvesting. Hemp is an important crop, used in the construction of fishing nets and basketry (Manila hemp or abaca).

Fishing is a way of life along the seacoast and on the larger lakes. Fish constitute a significant proportion of the protein in the Filipino diet. Filipinos have a rich tradition of boat building. Seagoing craft include outriggers with large sails. The lakes and rivers were navigated with various types of canoes. Open-air fish markets were prevalent in Filipino communities during the War.

Carabao or water buffalo are extremely important to the Filipinos. The domesticated animals, with the stocky build of a beast of burden, are used for many tasks. The animals assist with work in the rice paddies, plow fields, carry loads, and move heavy objects. They are important means of transportation in the Philippines, pulling wagons, sleds, and carriages. They can be ridden like a horse through all terrain, as they are comfortable in navigating through swamps, rivers, and mud. Carabao milk is rich and creamy, and the meat like beef, but leaner. The carabao cools itself by lying in a waterhole or mud "wallow" during the heat of the day. It feeds upon the abundant grasses and reeds growing in the wetlands. Mud, raked up with its horns and caked on its body, protects the animal from insects. Native to the Philippines, the carabao symbolizes strength, power, efficiency, and perseverance to the Filipinos. A family's carabao is a prized possession and very important to their livelihood. During the War over 70% of the Philippines' carabao population was killed. They were slaughtered for meat by the Japanese, and many were killed in the course of battle.

The climate of the Philippines is tropical and monsoonal. In general, rain-bearing winds blow from the southwest from approximately May to October, and drier winds come from the northeast from November to February. Thus, temperatures remain relatively constant from north to south during the year, and there are wet and dry seasons. The dry season generally begins in December and ends in May, the first three months being cool and the second three hot; the rest of the year constitutes the wet season. During the wet season, rainfall is heavy across the island chain. From June to December tropical cyclones (typhoons) often strike the Philippines. Typhoons are heaviest in Samar, Leyte, south-central Luzon, and the Batan Islands, and, when accompanied by floods or high winds, have been known to cause great loss of life and property. During the hot part of the dry season the temperature sometimes rises as high as 100 °F. Overall temperatures decline with elevation, however, and cities and towns located at higher elevations experience a pleasant climate throughout the year; the temperature may drop close to 40° F.

The U.S. in the Philippines

General Douglas MacArthur had a long history with the people of the Philippines. He had spent a considerable amount of his life there. His father had served there during the Spanish-American War and was acting

Governor-General of the U.S. Territory. In 1935, the President of the Philippines (Manuel Quezon) asked MacArthur to supervise the creation of the Philippine Army. Quezon and MacArthur had been personal friends for many years. With President Roosevelt's approval, MacArthur accepted the assignment. Prior to the outbreak of World War II, MacArthur was designated Field Marshal, Philippine Army, and acted as Military Advisor to the Commonwealth Government of the Philippines. In 1937, MacArthur officially retired from the Army. He would no longer represent the U.S. as military adviser to the government, but remained as Quezon's adviser in a civilian capacity. In July of 1941, President Roosevelt federalized the Philippine Army and recalled MacArthur to active duty in the U.S. Army, naming him commander of U.S. Army Forces in the Far East (USAFFE).

At roughly the same time as the Pearl Harbor attack, the Japanese invaded the strategically-important Philippine Islands. When the Japanese attacked the Philippines on December 8, 1941, American fighter aircraft were on patrol, but a ground fog delayed the Japanese aircraft on Formosa. When the attack finally came, most of the U.S. Air Corps was caught on the ground and destroyed by Japanese bombers. On the same day, the Japanese invaded several locations in northern Luzon and advanced rapidly southward toward Manila, capital and largest city of the Philippines. The U.S. army, consisting of both Americans and Filipinos, retreated onto the Bataan Peninsula. On December 26, 1941, Manila was declared an open city and American military forces abandoned the city leaving civilians behind. On January 2, 1942, Japanese forces entered and occupied Manila. They ordered all Americans and British citizens to remain in their homes until they could be registered. On January 5th, the Japanese published a warning in the Manila newspapers. "Anyone who inflicts, or attempts to inflict, an injury upon Japanese soldiers or individuals shall be shot to death." But if the assailant could not be found the Japanese "would hold ten influential persons as hostages."

The forces that MacArthur trained and equipped were no match for the Japanese. From the Bataan peninsula, they retreated to the island of Corregidor. MacArthur claimed that he wanted to fight to the end, but he was eventually ordered by President Roosevelt to leave. He, his wife, and young son Arthur had a harrowing escape. First by submarine, then by plane, they eluded the Japanese and arrived in Melbourne, Australia, on March 21, 1942. MacArthur left behind U.S. General Jonathan Wainwright, Major General Edward P. King, Jr., 12,000 American troops, and 60,000 Filipino troops. The troops left behind were expected to hold out until reinforcements could be sent. Unfortunately for them, it would be over two years until the U.S. returned.

On April 9, 1942, the forces on Bataan surrendered to the Japanese, against MacArthur's orders. The largest contingent of U.S. soldiers ever to surrender was taken captive by the Japanese. American and Filipino troops were led on the infamous "Bataan Death March," where 1,000 Americans and 9,000 Filipinos ultimately died under brutal Japanese treatment. Many of the P.O.W.s who made it alive to the Japanese camp subsequently died of starvation and disease.

Safely on Australian soil, MacArthur made his bold promise to the people of the Philippines, declaring, "I shall return." This became a rallying cry to the U.S. troops and the Filipinos who would eventually carry out his promise. News of the Bataan Death March did not reach the American public until January 27, 1944. Accounts of the atrocities committed by the Japanese caused outrage in the U.S. and increased public support for the war effort.

The Japanese in the Philippines

The Japanese occupation of the Philippines began on soon after they invaded in December of 1941. It would last for three years, from 1942 to the Japanese surrender in 1945. It was a cruel and bitter period for the Filipinos. Many Filipinos were forced into submission through fear, torture, and threat of death to family members. Over 1,000 young girls and women in the Philippines were sexually enslaved by the Japanese during World War II. Japanese treatment of the population got much worse as the War started to turn in favor to the Allies. The Philippines suffered great loss of life and massive physical destruction by the time the war was over. An estimated 527,000 Filipinos, both military and civilians, had been killed during the Japanese occupation period and the War. Of these approximately 150,000 were killed in seventy-two war crime events.

The U.S. controlled the Philippines at the time of the Japanese invasion and possessed important military bases there. Even though the combined American-Filipino army was defeated in the Battle of Bataan and the Battle of Corregidor in April 1942, guerrilla resistance against the Japanese continued throughout the War. Un-captured Filipino army units, a communist insurgency, and supporting American agents all played a role in the resistance. Due to the large number of islands, the Japanese never occupied many of the smaller minor islands. Japanese control over the countryside and smaller towns was also tenuous at best.

An active Filipino guerrilla resistance movement harassed the Japanese occupiers before the Americans returned. Prior to the invasion, the U.S. had set up relations with some of these Filipino guerrilla units. They were valuable in understanding how to navigate across the island. They also provided intelligence on the positions and movement of Japanese troops. Some Americans that escaped from Bataan also set up resistance units, some even with medical detachments. Their main purpose was not to engage the enemy in direct combat, but rather to gather intelligence for the planned Allied invasion.

Various rebel groups in the Visayas, the central islands of the Philippines where Leyte is located, worked with varying degrees of coordination with U.S. forces. One group, the Black Army, led by Ruperto Kangleon, played a crucial role in supporting U.S. operations, especially MacArthur's invasion of Leyte Island and the surrounding area. The guerrilla units armed themselves with any weapons they could find. Some rifles were recovered from the Japanese, others they made out of gas pipe. The gas pipe "guns" were loaded with gunpowder and old nails. The weapons were as deadly as any shotgun. The home-made guns were called "latongs" in the Visayan dialect of the Central Philippines, and "paltiks" in the Tagalog dialect of the Manila area. In some areas of Manila, residences were guarded since thieves would strip a house of gas pipe for the illicit shotgun industry.

The Hukbalahap, translated as "The Nation's Army Against the Japanese" was a socialist/communist guerrilla movement formed by the farmers of Central Luzon. They are popularly known as the "Huks." They were originally formed to fight the Japanese, but extended their fight into a rebellion against the Philippine government, known as the Hukbalahap Rebellion, in 1946. The revolt was later put down through a series of reforms and military victories. The Hukhbalahap's methods were often portrayed by other guerrilla leaders as terrorist; for example, Ray C. Hunt, an American who led his own band of 3000 guerrillas, said of the Hukbalahap that:

> *My experiences with the Huks were always unpleasant. Those I knew were much better assassins than soldiers. Tightly disciplined and led by fanatics, they murdered some Filipino landlords and drove others off to the comparative safety of Manila. They were not above plundering and torturing ordinary Filipinos, and they were treacherous enemies of all other guerrillas (on Luzon).*

However, the Hukbalahap claimed that it extended its guerrilla warfare campaign for over a decade merely in search of recognition as World War II freedom fighters and former American and Filipino allies who deserved a share of war reparations. After the war ended in 1945, the group was disbanded and vilified for its involvement in some of the Japanese atrocities in the islands. Individual members faced trials for treason as a result.

Some Filipino civilians assisted the Japanese, whether through threat, brainwashing, or believing that the Japanese cause would prevail. The Japanese had ways of persuading resistant locals into submission, like giving them scalding hot baths and freezing cold baths alternately, with no rest, no food, and no water, except the soapy water in the baths. Some may have been persuaded, as the Japanese soldier was, that dying for the Emperor would be the highest honor and guarantee eternal bliss.

The Makapili was a militant group formed in the Philippines in 1944 during World War II to give military aid to the Imperial Japanese Army. The group was meant to be on equal basis with the Japanese Army, and its leaders were appointed with ranks that were equal to their Japanese counterparts. It attracted approximately 6,000 members, many of them poor or landless farmers who came to the group due to vague promises of land reform after the war. The Makapili was not used to fight against U.S. forces but was deployed to counter the Filipino guerrilla units and Philippine military, particularly in rural areas.

October 20, 1944 – The Leyte Invasion

The Allied invasion of Leyte began on October 20, 1944. It was one of the largest battle armadas to ever be assembled. General MacArthur was designated the Supreme Commander of all sea, air, and land forces. Allied naval forces consisted primarily of the U.S. Seventh Fleet, commanded by Vice Admiral Thomas C. Kinkaid. Over 700 ships, including 157 warships, transported and put the landing force ashore. The U.S. Sixth Army, commanded by Lieutenant General Walter Krueger, consisting of two corps of two divisions each, conducted land operations. Krueger would land over 200,000 ground troops to face an estimated 85,000 Japanese. The Leyte invasion would be one of the greatest logistical challenges of the Pacific War. The invasion required unprecedented coordination between land, sea, and air forces. The U.S. supply line stretched for thousands of miles.

At the onset of the landings, the Navy's destroyers and battleships bombarded enemy positions on the island with their big guns and rocket launchers. A battalion of Army Rangers secured outlying islands and guided naval forces to the landing beaches. The new Sixth Army Service Command (ASCOM), commanded by Maj. Gen. Hugh J. Casey, was responsible for organizing the beachhead to supply units onshore, and constructing or improving roads and airfields.

Air support for the Leyte operation would be provided by the Seventh Fleet during the transport and amphibious phases, then transferred to Allied Air Forces, commanded by Lt. Gen. George C. Kenney, when conditions allowed. More distant-covering air support would be provided by the four fast carrier task forces of Admiral Halsey's Third Fleet, whose operations would remain under overall command of Admiral Nimitz. As it turned out, U.S. carrier-based aircraft would be kept busy engaging the Japanese Naval fleet. Marine Air Group 24 also carried out close-air ground support missions in support of the Army.

Unlike the Germans at Normandy, the Japanese Army had been caught by surprise at Leyte. The Japanese expected the invasion to occur on the southern island of Mindanao. Due to a lack of time and manpower the Japanese did not build strong defensive positions on the landing beaches. They had originally planned to fight the "decisive battle" for the Philippines on the northern island of Luzon. Instead of defending the beaches, General Yamashita planned to draw the invaders inland into the Central Valley and rugged Central Mountains. The mountains provided a steep, rocky fortification. From the heights the Japanese could look down into the flat, wet valley, raining fire upon the approaching U.S. infantry. The Japanese liked to fight from entrenched positions, like caves, pillboxes, and spider holes. Near the beaches and in the valleys, they hastily set up a network of coconut log pillboxes and spider holes to ambush the advancing troops. The Japanese also depended upon the environment to assist them. Rice paddies, swamps, and muddy roads would slow the advance of the U.S. invaders. Trucks and tanks would get mired in mud or have difficulty crossing water in the swamps, sometimes five feet deep. Narrow trails and thick jungle growth in the mountains limited visibility and would force the U.S. infantry to break up and advance in smaller units.

The Japanese kamikaze suicide planes made their deadly debut on Leyte. This desperate tactic would come as a surprise, even to those who had already witnessed Japanese brutality in many forms.

Gen. MacArthur waded through waist-deep sea water to the beach at Palo. His landing craft had been jammed up in the surf as the huge armada was unloading. The U.S. Navy CPO directing the landing craft, worn out and not impressed with the general, said "Let them walk." Japanese sniper and mortar fire were still active in the beach area. MacArthur, Philippine

President Osmena, and military staff stepped into the thigh-high water. It's been said that the general was fuming as he waded in through the surf. He was stiff-jawed and stared intently ahead through his Ray-Ban Aviators. Others seemed to walk behind him, maybe to avoid his angst, or just not wanting to be too close in case a sniper had him in their crosshairs. At that point an Army photographer snapped some the most dramatic and iconic images of the War. Always one for publicity, MacArthur ended up being very pleased with the photos. His dramatic landing hit the newswires around the world.

From the beach on Hollandia, Dad watched troop and supply ships depart for Leyte. Large flights of long-range bombers with fighter escorts assembled in V-formations. All made their way to join the invasion of the Philippine island.

The 44th anxiously listened for news of the invasion. Over the short-wave radio, they listened to MacArthur's dramatic speech. On the steps of Leyte's Provincial Capital in Tacloban, with the eyes of the world focused on him, MacArthur proudly announced, "People of the Philippines, I have returned."

Hopeful optimism was felt by those about to join the invasion force. News of the successful landing and Japanese defeat at sea brightened the spirits of the 44th. The U.S. seized territory on Leyte more quickly than expected. It seemed that the tide of the War was turning; the Allies were now on the offensive. But much territory still lay ahead, including the Philippine island of Luzon, the Ryukyus Islands, and the Japanese home islands themselves.

At the Granada theatre in Kansas City, Mom and her sister-in-law watched the newsreel of the invasion. Both wondered if their husbands were there. A few months before, other wives in the neighborhood had husbands that were part of the D-Day Invasion in Europe. Allied casualties in that invasion numbered over 10,000, with 4,414 confirmed dead. General MacArthur's confidence in the operation couldn't hide the fact that this was going to be a costly battle. Now it was Mom's turn to worry and wait.

The Japanese Imperial Navy did not plan to immediately attack the massive Allied landing flotilla. Instead they planned to wait until the troops came ashore. Then they planned a "decisive" naval battle with the U.S. fleet in Leyte Gulf. The IJN would encircle the U.S. fleet from the north and south. They would knock out the U.S. aircraft carriers and their planes, leaving the forces on the ground vulnerable to air attack. They would then fight a delaying battle, hoping to cut off the Allied supply line. It would be a battle of attrition where they hoped to prolong the war long enough to sue for a favorable peace settlement.

But, the Japanese hopes of crushing the U.S. naval fleet were dashed. The Battle of Leyte Gulf began on October 23rd. The battle is considered the largest naval battle of World War II and, by some criteria, possibly the largest naval battle in history, with over 200,000 naval personnel involved. A near-miss for the Allies, it turned out to be a devastating defeat for the Japanese. Their once powerful naval force would not be effective again. What was left of the Japanese Navy limped back to the north. There, they would repair and regroup for the inevitable defense of the home islands. The Japanese Navy would have little impact on the continued U.S. invasion of the Philippines.

November 11, 1944

The 44th assembled on the beach near Hollandia, New Guinea. The last of the hospital's supplies were loaded into a transport ship. On November 11, 1944, the 44th boarded the ship bound for Leyte. They joined a large armada of ships that departed from the port on the seven-day journey. Navy destroyers and recon planes escorted the ships. Moving across the large expanse of ocean under combat conditions was risky. Even hospital ships, though clearly marked with large crosses, were still being attacked. Japanese submarines patrolled the seas, protecting their shipping lanes to the southern islands. There was also the ever-present threat from the air as Japanese planes strafed, bombed, and torpedoed Allied ships.

As the men of the 44th sailed north towards the Philippine Sea, there was great anticipation of their landing on Leyte. Although they were not in the first wave of landings, the 44th felt the risk ahead of them. Once on land, the 44th would set up as a General Hospital. Expected to be at least four to five miles from the front lines they would treat the more serious casualties in the line of evacuation. Onboard ship, the doctors discussed the types of battle wounds expected. The administrative officers contemplated how they would transport the equipment and set up a 1,000-bed hospital facility. Supply officers pondered how basic food and water would be supplied in primitive conditions. Enlisted men rested while thinking of the work ahead of them. The 44th's training and preparation would now be applied. Some passed the time in card games. Some wrote letters home. All were told to prepare their wills. Tension started to build as the landing date approached. Some tried to lighten things up with humor. Some gathered in prayer and soul-searching. No one really knew what to expect. Would Japanese submarines attack the transport ships? Would Japanese aircraft harass the landing craft? What would the conditions on Leyte be like? But nothing they imagined would prepare them for what lay ahead.

As the ship travelled near the equator, the temperatures below deck were unbearable. Outdoor space on deck was preferred both day and night. On the decks, there was at least a breeze and a sky full of stars to help pass the night. At night, the ship's lights were blacked out to avoid observation. Dad said that he would look down from the top deck into the ship's wake. Long streams of phosphorescent marine life lit up the water in bright, sparkling colors. The plankton, fish and dolphins near the ship glowed brilliantly, an amazing sight.

The Philippines

I worked during my junior-year summer while school was out. It was a hot Friday afternoon and the clock was moving slow as molasses. Finally, it hit 5 p.m. The incessant drone of the factory's machines was replaced by laughter as another long work week ended and the day shift filed out. I parked the forklift I'd been driving. Covered in chalky chemical dust and sweat, I waited my turn at the time-clock, then punched out. In the hot asphalt parking lot some workers were already popping beers. Not me, not yet anyway. I headed home to shower, have dinner with Mom and Dad, and then meet my friends for a night out. I turned 17 in the summer of 1974. It was the "disco era." The local 18-year-old bars served watery 3.2% beer and played the latest disco until midnight. It was easy to get in with a modified driver's license and draft card. Back then your IDs were on paper. All you needed was an eraser and typewriter to instantly turn 18.

My high school friends Larry and Terry said that they'd pick me up after dinner. They always enjoyed talking to Dad and hearing his stories of the War. They also got a kick out of hearing Mom and Dad's humorous dialogs. But the last thing I wanted was to be at home on a Friday night. I knew that if they got Dad talking about the War, I could be stuck there for hours. My plan was to meet them in the driveway when they arrived. I'd tell them that Mom and Dad were getting ready to go to a movie. Then, we'd be on our way.

After dinner, I saw Dad sitting on the patio, so I went out to tell him that I was leaving.

"What are you doing tonight?" he asked.

"Oh, just meeting up with the guys," I replied.

"Need any cash?"

I couldn't refuse his generous offer of $10. In 1974 that was enough to pay for my beer, gas, and a couple of greasy Jack-in-the-Box tacos at the end of the night. I listened for the roar of Terry's black GTO convertible to come up our street. I suspected that my friends would be late as usual, so continued to sit on the patio with Dad. Suddenly I heard the side gate open, and my friend Terry announce, "Hey Mr. Odrowski, having a party tonight? I heard that there's going to be a lot of beer and BBQ!"

Dad chuckled, "Yes, you heard right." He called out to Mom in the kitchen, "Stella, get these boys a beer."

My plan to head them off had failed. I hadn't heard the car approach. Larry had driven tonight instead, in his Dad's quiet, 4-cylinder Plymouth Valiant. And, remarkably, they were actually on-time. It looked like my disco plans might be delayed.

Mom opened the screen door and came out with four Buds and a bowl of pretzels. She commented, "Ed, now don't keep these boys too long with your war stories. You know that the girls can't wait to see them tonight." That drew laughter from Terry and Larry. We all knew how far from the truth that was. Mom had a wry sense of humor. She and Dad played off each other like a 1970s sitcom couple.

Dad replied, "Stella, I've been telling these boys that they should join the service. When they get out, they'll be all grown up and even more popular with the girls."

Mom responded, "Now Ed, I don't think their mothers would want you telling them that!" She returned to the house to read the evening paper. It was getting late. I was anxious to go and tried to usher my friends out.

But then Terry asked, "So, Mr. Odrowski, what was it like to fight the Japanese?"

I popped the pull tab on my beer and settled in. I knew that we'd be here awhile.

Dad started in, "You know the Japs liked to fight at night. After dark we had to smoke in coffee cans. If you lit a match, a sniper would blow your head off. The 3rd anniversary of the attack on Pearl Harbor was coming up and we thought that the Japs might be up to something." Pausing to take a swig of beer, he continued, "Snipers had been firing into the camp at night. In the morning and evenings, I'd have my sergeants spray the tops of the nearby palm trees with their machine guns."

I'd heard this story many times before, so I played the straight man, asking, "How about during the day? Were you able to walk around in the open?"

Dad responded on cue, "Yeah, the Japs didn't like to fight during the day. It was the craziest thing. Here we were, in a war on a tropical island. Palm trees, beautiful beaches." He continued, "One evening I decided to go out and hit some golf balls." After a drink of beer, he grinned, "You know I had to keep up my golf game." The boys chuckled. In the 1970s this brought up an image of Colonel Blake from M*A*S*H, fishing hat and golf clubs, a medical unit in a war zone.

Dad continued, "I had emptied the hand grenades from my musette bag and filled it up with golf balls. As I was hitting balls out into the rice paddy, I heard the engines of the Jap planes. They came in low, just over the mountains, and flew right over our tents. Then I saw the paratroopers dropping, their chutes opening and men drifting down like leaves. About a dozen of them landed between me and the camp."

Terry eagerly asked, "What happened then? Did you shoot at them?"

"No, I reached in my bag for a grenade," Dad paused for effect, "But instead pulled out a golf ball!" Everyone laughed. I'd heard the punch line many times before, but loved it every time.

Larry asked, "Did they shoot at you?"

Dad continued, "No, strangest thing. They just collected their gear and joined the others that dropped on the airstrip behind us. I headed back to camp. It wasn't very long until all hell broke loose."

I might not ever know if the "golf ball" part of this story was true. But this story was the key thread that led me to discover the place, date, and historical context of the event. The members of the 44th experienced a series of unexpected events that placed them in a pivotal moment in the history of World War II. The courage and resolve of these "citizen" soldiers would be tested. In their first few months in the Philippines, they would be asked to perform well beyond their expected duties as a medical unit. The following sections describe their experience in the invasion of Leyte.

November 18, 1944

It was first light aboard ship. As the 44th's transport approached Leyte Gulf, the silhouettes of the small islands of Dinagat, Homonhon, and Suluan were seen on the horizon. A few weeks before, these were the first areas captured by U.S. Army Rangers at the start of the Leyte invasion. The seas were relatively calm. Tall, fluffy white clouds filled the sky. The members of the 44th felt the excitement of their upcoming landing. They were eager to get off the crowded ship, even if it meant walking into a combat zone.

As the larger island of Leyte came into view, they saw the U.S. Navy's battleships and destroyers lined up offshore. Their big guns were firing at inland targets. Like the 4th of July, this explosive display was dramatic in the pre-dawn hours. With deafening blasts and white smoke, the guns filled the air with the smell of sulphur. Rockets zipped out of launching tubes from smaller craft in an orderly fashion. As the sun came up, they could see black smoke rise in columns over the island, the remnants of shells that had reached their targets. Dad remembered the Navy's big guns as producing some of the most impressive sights and sounds of the War. He described the experience, "The blasts would reverberate inside of you. The shells roared like freight trains as they went overhead."

At that time, the Japanese still controlled the skies over Leyte and the landing beaches. Strafing and bombing runs continued to harass the landing and unloading of men and supplies. Land-based Japanese air power out of Luzon and Cebu outnumbered the Allies. Due to wet conditions and inadequate runways on Leyte the U.S. had still not established ground-based airfields. Plus, the Navy aircraft carriers were busy dealing with threats from the Japanese Naval Fleet off the coast of Leyte. There was a limited number of U.S. fighters, including P-38 Lightnings and the Navy's F6F Hellcats, that were available to engage the Japanese planes.

The men of the 44th had assigned duties onboard ship, one of those being "battle stations" in the event of an air or sea attack. Army personnel had been trained by the Navy gunners onboard to man anti-aircraft and machine guns. Army officers were assigned to direct a group of enlisted men at a machine gun post. Others were to assist with firefighting and care for the wounded, should an enemy attack be made.

As the 44th's transport ship entered Leyte Gulf, battle stations was sounded. The men quickly went to their positions. The look and demeanour of the sailors indicated that this was a real threat. In the distance, two planes were seen passing in and out of the high clouds. The Navy anti-aircraft guns opened up, filling the sky with black puffs of smoke. It was like a 4th of July "blockbuster" display. But the planes evaded the flack bursts. The men at the machine gun posts waited anxiously, the planes still out of range. As the planes flew into a clearing, they abruptly split apart. One veered to the right, the other dipped lower towards the sea. Two U.S. P-38 fighters were seen pursuing the lower plane from behind. The planes moved past the bow of the ship, low over the water. The machine gun posts were told to hold their fire. Now passing along the starboard side, the markings of the lead plane clearly identified it as a Japanese Zero. Looking on in awe, the men on board watched the chase. As the Zero tried to evade, a U.S. P-38 closed in from behind and fired its machine guns. A small trail of white smoke was seen streaming from the enemy plane. The Zero then burst into flames and broke up into multiple fragments. Dropping like a shattered clay pigeon, its parts splashed into the sea. The men on board cheered wildly. The speed and agility of the P-38 would prove to be very effective against the Japanese aircraft.

With the immediate danger averted, the orders were given to board the landing craft. A Navy LST (Landing Ship, Tank) lined up on the port side of the ship. LST is the designation for ships first developed during World War II to support amphibious operations by carrying tanks, vehicles, cargo, and landing troops directly onto shore with no docks or piers. This enabled amphibious assaults on almost any beach in the Pacific. The bow of the LST had a large door that would open with a ramp for unloading the vehicles. The LST also had a special flat keel that allowed the ship to be beached and stay upright. The twin propellers and rudders provided protection from grounding. Troops cynically nicknamed the LST "Large Slow Target." Not fast movers, they were vulnerable to attack from the air.

It was a tricky manoeuvre to climb down the rope netting from the taller ship to load the LST. A large ocean swell could swing the net out and crash back into the ship's steel hull. Dad had already been injured once while unloading, when grounded on the reef on their way to New Guinea.

The Navy crew remained at battle stations while the LSTs were being loaded. The antiaircraft guns suddenly started firing again. Another plane came into view, approaching from directly above. The ship's gunners attempted to hit the diving plane with anti-aircraft flack and .50-caliber machine guns. Tracer bullets from the machine guns made a glowing path across the morning sky. The plane averted the fire, amazingly moving untouched through the black puffs of smoke that pockmarked the sky. The plane dove at a sharp angle and headed directly towards the transport ships. Onboard the LST, the men watched in horror as the plane crashed into a transport ship that had just unloaded, less than a hundred yards to the right of theirs. It smashed through the deck and exploded in the mess area where the crew members were having breakfast. Flames and black smoke were seen rising from the deck of the ship. The attack killed 135 U.S. Navy sailors below deck and injured many others. The bomb the suicide plane carried did not detonate or many more would have been killed. But the fuel remaining in its tanks had done its deadly work. A few hours later the blazing ship sank into the waters of Leyte Gulf. The 44th had witnessed, first-hand, the devastating effect of the kamikaze. The 44th's personnel were loaded into the landing craft. The Navy crew seemed eager to get rid of their Army passengers and head back out to sea.

The 44th's landing destination was White Beach, the northernmost landing beach near the coastal town of Tacloban, Leyte. White Beach isn't named for its white sand, it was the code name for the location. It has greyish sand and murky waters. A few miles to the south was Red Beach near Palo. This was where General MacArthur came ashore on October 20th. The water was much clearer there.

As the 44th's craft approached the landing beach, there was anticipation, apprehension, and outright fear. Many smoked a last cigarette as they anxiously waited. As they breathed in the grey diesel smoke and felt the motion of the wave swells, some of the men got sick. Anxiety may have contributed to the seasickness. As the LST stopped just short of the beach and lowered its ramp, the men drew nervous breaths. The kamikaze attack they had witnessed offshore was a chilling reminder of their vulnerability. The sights, sounds, and smells of the landing would not be forgotten.

The large gangway of the LST opened into the shallow surf. It looked like a beached whale that opened its large mouth to disgorge its stomach onto the beach. The men grabbed their packs and waded ashore in the surf, water up to their knees. Once on the beach, they moved in an orderly fashion to an area of downed coconut logs. Next, their supplies were unloaded from the LST. The beach area seemed safe, with no enemy gunfire nearby. There was only the sound of U.S. artillery shells flying overhead that were directed at inland targets.

Then two planes were seen on the horizon further up the coast. As the planes moved closer, the men were told to take cover. A Japanese Zero arrived first and began strafing the beach with machine gun fire. The other plane, a Mitsubishi "Betty" bomber, flew in low after the Zero. It released a bomb that fell through the sky at a sharp downward angle. The men watched in shock as the bomb made a direct hit on the LST they had just come ashore on. Captain Willard Arnquist, MD, recalled that second close call, "We felt relieved to make it to the beach. Then a Japanese plane came in low and dropped a bomb on the LST that had just finished unloading our supplies. Fortunately, it was a dud."

Upon gathering their supplies, the 44th immediately went into action as a medical unit. The dead and wounded from the kamikaze attack on the transport offshore started arriving in small landing craft. The 44th's training and teamwork enabled them to set up a makeshift hospital on the beach within hours. The kamikaze inflicted horrific wounds in its use during the remainder of the War. Many of those wounded in attacks would suffer burns over a large percentage of their bodies. Walter Teague recalled, "We removed 135 charred bodies to the beach for burial. That was our welcome to the Philippines." Survivors, many burned significantly, were treated in the makeshift hospital on the beach.

In their first few hours, the 44th had experienced what combat would be like on Leyte. Even for the experienced medical professionals, it was shocking to see this type of carnage. There would be much more to come as the fighting on Leyte raged. It was sobering to note that the wounded or killed in combat are not always those on the front lines of battle. They also include supply personnel unloading cargo, the cooks preparing meals, the dishwashers cleaning up, and in the case of the kamikaze attack, those just sitting down to have breakfast.

With their landing on November 18, 1944, at White Beach, Leyte, the 44th had become part of a significant historic event. They were participants in the largest invasion of the Pacific War, one that would prove to be a major turning point in the war against Japan. In an official U.S. Army medical journal, it was noted that unlike other hospital units that arrived later, the 44th landed on Leyte "with pup tents and K-Rations." This was a testament to their ability and willingness to operate in tough conditions.

November 19, 1944

The 44th was assigned to the village of Burauen, approximately 20 miles inland from the landing beach. The 6th Army's 7th Infantry Division had just cleared the area days before. The battle for Leyte raged in the inland valleys and foothills of the Central Mountains near Burauen. Hospital staff and medical supplies were in great need as battle casualties increased.

The village of Burauen was at a strategic central crossroads on the island of Leyte. The peaks of the Central Mountain range covered the horizon to the west. The name evolved from the Visayan name Buraburon, meaning "having many springs." Both cold and hot springs are abundant in the area. A number of rivers and lakes formed from the springs and the runoff from the mountains. Geothermal activity is evident and associated with the nearby Mahagnao volcano. The village was surrounded by swamps, rice paddies, and lush groves of coconut palms. In the mountains, steep, narrow trails led from Burauen through dense jungle growth. The trails connected Burauen with a number of smaller mountain villages, and eventually to Ormoc Bay on the west coast.

Burauen was a village of approximately 30,000 inhabitants that lived in crowded rows of homes made of wood and tin with palm-thatched roofs. The natives were mainly farmers and fisherman who worked in the nearby fields, rice paddies, lakes, and rivers. Tropical fruits including bananas, mangoes, papaya, and coconuts were harvested locally. Camote (tasting like a sweet potato), beans, and other vegetables could be found in family garden plots. In front of their thatched homes, Filipino women pounded rice as they had for centuries. Seafood was brought into the markets from the east coast. Farmers raised chickens for eggs and meat. The fittest roosters were selected to compete in the favorite local sport, cockfighting. Cockfights were held every Sunday afternoon and included wagering and local libations.

Japanese occupation had been brutal at Burauen. Thousands of villagers were killed and many properties destroyed. The sight of orphaned children, many undernourished, provided a grim reminder of the impact that war has on the innocent. Filipino women, some just young girls, were abused by the Japanese. Many were detained as "comfort women," the Japanese euphemism for sex slaves. These women provided sex for the Japanese officers and occupying troops. Some were raped over 20 times per day.

Since the Japanese occupation, the Filipino guerrillas clashed with sympathizers in the village. Now that the Japanese retreated, there was swift retaliation on those who supported the Japanese. The Mayor of Burauen, a Japanese sympathizer, had spread propaganda promising a better future with the Japanese. Upon the U.S. infantry arrival, he was taken down to the river by Filipino guerrillas, not to be seen again. Apparently, he was not a good swimmer.

Close to the village of Burauen, the Japanese had built three airfields, named Buri, San Pablo, and Bayug. They were located within a mile of each other. The Japanese used the airfields to support their land-based fighters and bombers. By October 27th, the U.S. Army's 7th Infantry division took control of Burauen village and the airfields. An Army engineering unit attempted to repair the airfields. The U.S. 5th Army Air Corps started operating fighters from the facility. Bayug became the home for the 110th Reconnaissance Squadron's P-40s on November 3rd. P-38 fighters started using Buri on November 5th. But this didn't go well from the start. A major typhoon had hit Leyte on November 9th. The rainy season in the Philippines was in full swing. Over 35 inches of rain would fall in just a few weeks. The rain would have a profound impact on Allied ground and air operations. The soggy, unstable runways made the airfields unsuitable for the fighter groups. Pilots had difficulties operating the heavier U.S. planes there. A number of P-38s and other U.S. planes lay wrecked along the runways.

Already behind schedule, General Kenney, head of the 5th Army Air Corps, decided to abandon the Buri Airfield. U.S. air and ground units also started to pull out of the San Pablo Airfield. This left the Burauen area without an active fighter group, but Bayug and San Pablo continued to operate L-4 liaison aircraft. The single-engine L-4s (i.e., "Piper Cubs") were used as recon planes, for evacuation of wounded, and to airlift supplies. The 5th Army Air Corps kept their headquarters at Burauen. But the only Air Corps personnel left consisted of the L-4 pilots, mechanical staff, and other service personnel. Army Engineers established more suitable airfields along the less swampy east coast, at Dulag and Tacloban. By November 19th, U.S. fighters began to operate there.

Due to the heavy rains, access from the narrow landing beaches was blocked by swollen rivers and seemingly bottomless swamps. Vehicles were not able to leave the beach "en masse" until the engineers constructed roads to support their weight. The 127th Airborne Engineer Battalion was tasked with this work. Army Engineering equipment and support for setting up hospitals was limited. Construction resources were allocated to the repair of roads, bridges, and airfields, all essential to support the infantry in the invasion.

Hospitals, like the 44th, faced a logistical nightmare. An Army General Hospital required a caravan of trucks to haul the tents, personnel, and supplies to its designated area. Construction equipment was needed to clear an area and set up water supply and sanitation. Due to shortages, the 44th had to scrounge for equipment and vehicles. They would need to hire local laborers or perform the construction work themselves. This was a widespread problem for medical units at the time. Corpsmen and experienced doctors would have to pitch in on hospital construction projects. As well as treating the wounded, surgeons were also tasked with mixing concrete, sawing boards, and hammering nails.

Dad and other 44th officers were sent ahead to Burauen, while the rest of the 44th stayed on the beach to gather men and supplies into a large caravan. The officers planned to meet with other Army units and local officials in the Burauen area. Accompanied by members of an infantry unit, they loaded into jeeps and started the trek inland. On the journey inland, the 44th's officers encountered the bad roads and thick mud caused by heavy rains. Death and destruction were everywhere along the slow 10-mile route. Dead Japanese were seen on the roadways and laying in the ditches. There were also rotting carcasses of animals that had been killed in the fighting. Fleeing Filipino men, women, and children, many displaced from their homes, walked or rode in carts toward the landing beaches. Fearing retaliation by the Japanese, they sought the protection of U.S. troops. Also, many had suffered injury in the fighting and sought medical attention.

As a young boy, I liked to watch the many World War II movies made in the 50s and 60s. The classics included *From Here to Eternity, The Bridge over the River Kwai, The Longest Day,* and *The Great Escape.* The stars included John Wayne, Jimmy Stewart, Steve McQueen, Henry Fonda, and Robert Mitchum. Some of these icons had actually served in the War. The movies presented a clear dichotomy of the "good guys" vs. the "bad guys." The War was waged on the battlefield by powerful men, blasting it out, until one side stood victorious—usually the "good guys." Topics like civilian casualties were rarely covered, as the most popular World War II movies usually didn't deal in controversial subjects.

But in the late 60s, at the height of the Vietnam War, the evening news reports were showing a different side of war. Graphic reports of civilian casualties were covered. These included stories of atrocities, allegedly committed by both sides, in the conflict. The situations encountered in war seemed very complex. How do you know who to trust? How do you know how to react in a possible life or death situation? This led me to wonder about Dad's experience in the War.

One evening on the patio, I asked Dad a direct question, "Did you see civilians killed during the War?"

He replied, "Yes, civilians were killed by bombing and many in the Philippines were executed by the Japanese."

I could tell that this was a touchy subject to probe into.

"Why are you asking?" he said.

I had heard of the horror of the holocaust in Europe, the Japanese conquest of China, and the Allied bombing of European and Japanese cities. Millions of civilians had been killed from these devastating events. But that wasn't what I was asking about. I felt uncomfortable asking, but got to the point, "I mean, did our troops ever shoot civilians?"

I explained that I had been watching the evening news and saw stories about Vietnamese villagers being killed. At 12 years old, this was not my "image" of what World War II was like from the old newsreels and movies.

I could tell that this was a disturbing part of the War for Dad. With the other stories he was usually very animated. But this subject was sobering. He continued, "Some of the other officers and I were sent inland to prepare a site for the hospital and to meet the village officials. We needed to find Filipino laborers that we could hire to help with construction tasks. I selected some of my sergeants and enlisted men to go. We were guided by an infantry platoon."

Dad continued, "As we approached one village we were stopped by a young corporal."

"He told us that the area was too dangerous to enter. Their lieutenant advised that we waited until it was clear. He said that they had found some Japanese soldiers held up in some village huts. They had a firefight before the Japanese took off. But the worse part was yet to come. A woman from the village came up to one of our men. She was carrying a baby. He approached her, not expecting anything. When she got up to him, she dropped two grenades at their feet and they were all killed."

Dad continued, "We waited on the edge of the village until we got the all clear to move through. As we walked through, we saw a few huts that were still burning. I saw the bodies lying on the road as we passed by. It was a terrible sight. Moving inland through the villages we had to always be on alert. GIs in the lead would sometimes have to point their rifles and order people to stay back. There were tough decisions that had to be made."

This was confusing to me. In war, how do you know who you can trust? At that time, I couldn't comprehend why someone would go the extreme of attaching a grenade to their belts, pulling the pin, and dropping it to kill themselves, their baby, and a soldier. The killing of civilians out of anger was definitely wrong, but I now understood that there were situations faced that posed the choice, kill or possibly be killed. I had a different perspective when I watched the news from Vietnam.

Dad said, "Then I saw the craziest thing."

Not knowing what to expect next, I wasn't sure I wanted to hear the story.

But Dad continued, "As we walked through the village, a dog came out from behind a hut and ran at one of the GIs in the lead. He opened fire before it could get close."

"Shot the dog?" I asked.

"Yes"

"Because it was going to bite him?"

"Worse than that. The dog had a grenade rigged to its jaw with a wire attached to the pin. When it opened its jaw to bite, the grenade dropped. Luckily this one was a dud."

Later in life, it was clearer to me how these extreme aspects of war could forever leave a scar on someone's psyche. Also, that war wasn't like it was depicted in the popular movies of my youth.

Upon arriving at Burauen, the officers saw the devastation, including scores of orphaned children. In their faces they saw the impact of sickness, hunger, and despair. But the villagers also expressed their relief in seeing the Americans arrive and drive off the Japanese. Dad and the officers met with local officials to arrange native laborers to assist with the task of setting up the hospital. The villagers were eager to help in the cause and also make some needed money.

November 21, 1944

Back on White Beach, the 44th staff worked day and night to unload and organize their equipment and supplies for the move inland. At night they worked under floodlights. These would have to be turned off each time a Japanese recon plane approached. The men became familiar with a plane they called "Washing Machine Charley," aptly named because of the churning sound made by its misaligned Mitsubishi engines. The planes would circle, drop a few bombs, then head back to their base.

The Japanese continued to take advantage of their undisputed control of the air over Leyte. Every morning and evening, Japanese planes, including Zeros and Betties, would strafe and bomb the beach. They would approach low and from the east in the morning, and low and from the west in the evening. With the sun behind them it limited the vision of U.S. anti-aircraft gunners. The men of the 44th soon learned that a foxhole dug near the beach would fill up with water. During Japanese strafing runs they instead sought the protection of large coconut palm trees and downed logs away from the beach.

A Japanese Bettie bomber arrived in the evening from its base on Cebu. The plane approached from the west. Hearing the approaching plane, the men of the 44th took cover in a grove of coconut palms. The bomber released a "broomstick" bomb into the grove. When the bomb exploded, Lieutenant Joe Young, the 44th General Hospital's pharmacist, was hit in the back as he took shelter near a tree. The shrapnel severed his spine. Although quickly attended to by the doctors close by, he passed away within an hour. Broomstick bombs were a deadly weapon. Also known as "daisy cutters," they had an attached pole, the "broomstick," that would cause the bomb to explode before it hit the ground. Its shrapnel would spread above ground in all directions, hence the term "daisy cutter."

The 44th now dealt with the loss of one of their own, a friend, colleague, and key member of the medical team. He was a doctor and fellow soldier, someone they had shared memories with back home as they prepared for the War. A fellow traveller on their journey overseas, not to return to a life in the States. The loss weighed heavily on everyone. It was a sobering reminder that this could happen to any one of them. You could be treating the wounded one minute, the next minute this could be your fate. The 44th's officers collected Joe's personal belongings to send back home. They wrote letters to tell the family of his service and the great sacrifice he made.

On November 21st the 44th received orders to leave White Beach and depart for Burauen. The realities of war had already made a big impact on the unit.

November 22, 1944

When the 44th's caravan arrived at Burauen, Filipino laborers were already hard at work. A flat area about a mile north of the village was selected as the hospital's site. It was close to a main north-south Dagami-Burauen Road and the Buri Airfield. Filipino laborers sawed down trees, cleared brush, repaired roads, and created drainage ditches. The laborers helped pitch the large hospital tents. Much of the hospital setup would have to be accomplished without the benefit of heavy equipment and Army construction workers. Engineering problems on Leyte were much greater than anticipated due to the heavy rains. The most urgent need for Army engineering units was devoted to work on repair of airfields, roads, and bridges. The shortage of construction equipment was partly due to a supply logjam offshore. Equipment was stuck in the harbor for lack of transport to the landing beach.

The corpsmen and doctors of the 44th also pitched in as construction workers alongside the Filipino laborers. Setting up the hospital included establishing a water supply and managing sewage and sanitation. Mosquito control was essential to prevent malaria. But this would be next to impossible as the hospital was within yards of a large mosquito-infested swamp. Due to the heavy rains, drainage problems were a challenge. The hospital staff used wooden boards from every spare crate to place under the legs of operating tables and cots to keep them from sinking in mud. Walkways through the mud were also lined with any material available, including metal tracks removed from the airfields the Japanese had built.

Amazingly, a 1,000+ bed hospital was set up and operational in a matter of a few days. Very few hospitals were operating inland on Leyte. Many were still stuck on the beach waiting for their supplies. Others were limited in their movement by the impassable roads and shifting battle fronts. Originally intended to be a typical General Hospital located many miles from the front lines, the 44th now found themselves very close to the fighting. The sounds of battle raged day and night and Japanese snipers were active in the area.

The 7th Infantry Division, which had been securing the Burauen area, was ordered to move in force to the west coast. There at Ormoc, they would meet up with a U.S. landing force and push the Japanese north. On November 22nd General Krueger ordered the 11th Airborne Division to relieve the 7th Infantry at Burauen. The 11th Airborne Division had landed on Leyte on November 18th at Bito Beach, near Dulag. This was the southernmost landing area, also known as Yellow Beach. Originally, they were assigned to Leyte to stage for the planned invasion of Luzon. But the shifting tides of battle changed that. The 11th Airborne would now assemble at Burauen. The 11th's commander, General Joseph "Jumpin' Joe" Swing, set up his headquarters at San Pablo by the airstrip of the same name. Swing was a classmate of General Dwight D. Eisenhower, and also a member of the Army football team with him.

General Swing was directed to block Japanese movement through the Central Mountains to the east. One of the 11th's battalions set up a base near a large Filipino guerrilla camp in the mountains. Added to their mission was the protection of Army Air Force areas to the rear at Burauen. With this order, they would become closely aligned with the 44th. Smaller than a standard infantry division, they numbered 8,000 men. The airborne unit was also more lightly armed than regular infantry. In great physical condition, they were intended to move swiftly in small groups and operate in difficult terrain. They were well-suited for their operation in the Central Mountains. But if met with heavy enemy resistance, they were not as well supported as a regular infantry division. Swing continued to shift more of his troops to the mountains west of Burauen as encounters with the Japanese increased.

Evacuation of the wounded from the Central Mountains was difficult and slow. General Swing determined that he could use L-4 single-engine planes to evacuate his 11th Airborne wounded by air. The helicopter, which became key for evacuation of wounded in the Korean and Vietnam wars, had not yet come into widespread use. Swing ordered the 127th Airborne Engineer Battalion to jump from small planes with their construction gear and clear an area for a landing strip on the top of a mountain. While the strip was being constructed, the 221st Airborne Medical Company dropped surgical teams and technicians into the area to immediately treat casualties. This was another example of how the Army went to great extents to evacuate wounded as quickly as possible.

L-4 liaison planes operating from the Burauen airfields also dropped supplies by parachute to reinforce the 11th Airborne troops in the mountains. Rod Serling, the creator of the TV series *The Twilight Zone*, was a young member of the 11th Airborne fighting on Leyte. He and a good friend would perform skits to entertain the men in their unit. During one of the supply drops, his friend was joking around about what they were going to receive in the drop. He hadn't noticed a large crate dropping by parachute overhead. As the crate arrived, it struck and killed him instantly. It was said by Serling's biographers, that the horrific events and experiences from the fighting on Leyte greatly influenced his future writing. His themes centered on the macabre and absurd. He suffered from nightmares and flashbacks for the rest of his life. Serling once said, "I was bitter about everything and at loose ends when I got out of the service. I think I turned to writing to get it off my chest."

November 23, 1944

Helmuth von Moltke, a 19th century Prussian Field Marshall, has been quoted as saying "No battle plan survives first contact with the enemy." This was true for the Japanese at Leyte. Things were not going well for them in the month following MacArthur's landing. The U.S. infantry had shown their firepower and tenacity in systematically breaking up fortified Japanese defenses. It appeared inevitable that, if not held in check, the U.S.

73

6th Army would push to the west coast and the port of Ormoc. With that, the Japanese reinforcement and supply line would be cut and their infantry isolated. Since the Japanese Navy had been soundly defeated in the Battle of Leyte Gulf, they wouldn't be able to stop the U.S. from landing additional troops and supplies. If Leyte fell, Luzon, to the north, would be next.

The shifting tides of battle caused the Japanese to rethink their strategy. Generals Yamashita and Suzuki decided on a bold new plan. Leyte would now be the stage for the "decisive battle" they had hoped for, and not Luzon. They would launch a coordinated airborne and infantry counterattack on Leyte. Their attention focused on the central crossroads of Burauen. Their initial objective was to take back the Burauen airfields. The airfields were of high importance to the Japanese, seen as critical to restoring their air power on Leyte. The need for the counterattack was significant for the Japanese. Their troops and civilians were facing starvation as supplies of food, oil, and other necessities were being cut off. Japanese reinforcements and supplies were still flowing in from the west coast of Leyte. If the Japanese wanted to extend the war, they needed to maintain these supply routes. Once Japanese planes were back in operation at Burauen, they could safeguard their supply line and hold back the American advance. The risk was great for the Japanese. Suzuki risked losing his best infantry divisions in the process. But with their options running out, Yamashita and Suzuki were willing to gamble.

The Japanese counterattack on Burauen was planned for early December. The "Wa" operation would combine the Japanese 16th and 26th infantry divisions, already on Leyte, with airborne units from Luzon. General Suzuki's infantry divisions would assemble in the Central Mountains just west of Burauen and attack in unison. The soldiers of the reviled 16th Division had participated in both the Rape of Nanking and the Bataan Death March. The "Te" operation, would be the airborne part of the assault. Paratroopers of the 4th Airborne Raiding Division would assemble in Luzon. In a coordinated attack, the IJA's 2nd Parachute Group, composed of the 3rd and 4th Parachute regiments, would land on the three Burauen airfields. Airborne forces would also take back the Dulag and Tacloban airfields on Leyte's east coast. Allegedly, the Japanese did not know that U.S. fighter aircraft had abandoned the Burauen airfields. Unbeknownst to the 44th, they were soon to be at ground-zero of the planned Japanese counterattack on Leyte.

November 24, 1944

The 44th started taking in casualties on November 24th. Fierce fighting in the Central Valley and Central Mountains of Leyte sent a steady stream of wounded to the hospital. The rough terrain of the mountains and the swampy, muddy lowlands made the evacuation of wounded a challenge. Wounded arrived by truck or jeep, or by C-47 transports from evacuation hospitals. Casualties from the 11th Airborne, fighting in the mountains, were carried in on litters by Filipino laborers. Due to the rough terrain, it could take precious days to get wounded to the hospital. From the outset of the invasion 80% of casualties were battle-related and 20% medical. This ratio would shift the longer the troops were on the island. Medical cases due to battle fatigue, tropical diseases, and the wet, humid environment would later become more prevalent.

Mud and the problem of wet feet remained a serious problem at Burauen. One of the 44th's officers discovered a solution. He noticed that the Army's African-American truck drivers, who delivered supplies from the east coast, were better equipped for the environment. They wore knee-length rubber boots, well-suited for the muddy ground. The officer had a plan. He had stashed some whiskey from their less stressful days in Australia. As the medical officer described, "I stopped one driver, pointed to a bottle of booze and offered it to him for his boots. The driver went home in his stocking feet and I had boots for the duration. The next day the camp was overrun by truck drivers with boots in hand."

The mud had a more ominous impact on those caring for the wounded. Walter Teague described that many wounded died in the field while seeking help, "face down and alone in the mud." Desperately trying to reach a medic or get to an aid station, they crawled on their bellies before succumbing. Sometimes battle conditions inadvertently left wounded men behind. The medics of the 44th were committed to eliminating this most undignified way for a soldier to die.

November 27, 1944 – The First Japanese Paratrooper Drop

However unappetizing the packaged field rations were, American service-men could count on a "taste of home" on two major holidays, Thanksgiving and Christmas. For those fighting in the European Theatre in 1944, General Dwight Eisenhower ordered that every soldier under his command should receive a full Thanksgiving Day turkey dinner. A national Thanksgiving Day proclamation was issued by President Franklin D. Roosevelt on November 1, 1944. The National Day of Thanksgiving was to take place on November 23, 1944. Sometimes field and combat conditions necessitated a later celebration. Too busy during their move to Burauen, the 44th's feast had been delayed. But with mess facilities fully operational, Walter Teague, Mess Officer, could finally deliver the feast.

As the men celebrated the holiday, popular music of the day was being played from the radio over a large set of speakers. A Japanese station was the only source to hear the music of Glenn Miller, Benny Goodman, Harry James, The Andrews Sisters, and Bing Crosby. The music would be inter-spersed with Japanese propaganda reported by the announcer, called Tokyo Rose. Her reports, seldom factual, would exaggerate Japanese victories and warn of pending disaster for the Americans. Dad said that it was disturbing that Tokyo Rose would call out the U.S. units on Leyte by name, and even the names of some of the officers. The men would dismiss her reports and joke about them. But, while taking a break for the holiday meal, the members of the 44th were not aware of the mounting Japanese threat.

The Japanese Fourth Air Army determined to send an elite airborne unit to Burauen to inflict damage to the airfields prior to the main counterattack. The Kaoru Airborne Raiding Detachment assembled on Luzon. Kaoru translates to "distinguished service" in Japanese. The men of this unit were recruited from the Takasago tribe on the island of Formosa. They   76 were considered courageous and skilled jungle fighters who carried a tra-ditional knife called "giyuto" or "loyalty sword." Officers and other soldiers of the detachment, including medics, were from Japan. The paratroopers were trained at Tokyo's famed Nakano Intelligence School in guerrilla and infiltration tactics. The plan, called the "Gi" operation, was for some of the guerrillas to crash land on the Burauen airfields in Type 100 "Helen" heavy bombers. They were to use demolition charges to destroy enemy aircraft on the ground. Others would jump directly on the airfields from Type 100 "Topsy" transports and attack U.S. positions with rifles and machine guns. After their initial mission was accomplished, the unit planned to disperse into the jungle and carry out further guerrilla attacks on American positions.

On November 27th, the Kaoru Detachment assembled at Nipa Airfield on southern Luzon. As they prepared to depart, they raised a cheer to the Emperor, "Tenno haika! Banzai, banzai, banzai!" (may the Emperor live ten thousand years). Lieutenant General Kyoji Tominaga, commander of the Fourth Air Army, shook hands with the Kaoru Detachment's leader, Lieutenant Shigeo Naka, and wished the unit well.

Forty paratroopers from the Kaoru detachment left Luzon in four aircraft, ten men per plane. They proceeded on their flight to the Burauen Airfields, flying low to avoid detection. Two hours later the pilots reported back to the Japanese Fourth Air Army headquarters that they had reached their target. No further word was received from them. The Japanese assumed the mission was a success. But that was not the case. Most of the paratroopers ended up dropping on the wrong location. One plane landed in the sea near the Dulag airfields, far off target. Another crash landed on Bito Beach, near the Abuyog airfield. American troops, thinking that a U.S. C-47 had crash-landed raced to the scene in a jeep. Arriving at the scene, they were greeted with hand grenades. In the firefight that ensued, the U.S. troops there killed one raider. The other nine survivors vanished in the adjacent jungle. Another plane missed its target completely, landing near Ormoc on the west coast, where the raiders joined other Japanese ground troops. But one plane still appeared to be on target.

As some of the members of the 44th were finishing their Thanksgiving dinners, the hum of a plane was heard coming down the valley. Some standing outside thought that it was a U.S. C-47 transport plane bringing in supplies or wounded. As it got closer, they saw the markings of the Rising Sun. In an instant, a U.S. P-38 appeared and closed in quickly from behind. The men watched in awe as the Japanese plane was hit by fire from the P-38. The plane veered off to the west, crashed into a mountain, and burst into flames. The satchel bombs the paratroopers carried likely contributed to the massive explosion. All on board were killed.

In the paratrooper attack, all four Japanese planes failed to meet their objectives. But when few U.S. planes were seen over the west coast of Leyte the next day, the Japanese celebrated their supposed success. In a radio broadcast Tokyo Rose praised the success of the "fearsome" airborne raiders. She advised the U.S. troops on Leyte of the futility of their cause and that they should surrender. She warned that soon the Americans would be "gassed off the island." Although the paratrooper raid on November 27th had actually failed, the Japanese continued to believe that it was feasible to execute an airborne attack on the airfields. Their counterattack plans were now in full motion.

The Japanese airborne attack at Burauen gave the U.S. Army leadership some cause for concern, but it wasn't enough to distract them on their mission to push the Japanese off Leyte. General MacArthur was in a hurry to get to the west coast of Leyte. The Japanese continued to reinforce their troops and supplies at the west coast port of Ormoc, which would prolong the conflict on Leyte. The U.S. planned to land an additional invasion force on the west coast in December. They would cut off Japanese access to the port city of Ormoc and surround them on the island. By driving the Japanese off Leyte, Mac could stage for the real prize, the main Philippine island of Luzon to the north, and its major city, the "jewel" of Manila. Once Luzon was liberated, Mac would fulfill his promise to liberate the Philippines.

Japanese Infantry General Suzuki had no intention of letting the U.S. surround him. Instead, he seized the initiative and launched the planned counterattack. The attack was of utmost importance to the Japanese. The war continued to use up resources of food and fuel. If the U.S. took control of the Philippines and its air bases, they could control the flow of oil, rubber, and food from the Southeast Pacific. But if the Japanese could re-establish their airfields on Leyte, they could possibly prevent the invasion of Luzon and other Philippine islands. They could stall the American forces in a war of attrition and potentially negotiate a favorable peace deal.

General Yamashita emphasized this point to Suzuki. A message he sent stated that "the prevention of the establishment of American air bases on Leyte was critical not only to the Battle for Leyte, but to the continued well-being of Japan as well." He ordered the conquest of the airfields at Burauen, as well as those at Dulag and Tacloban. Simply put, "eliminate American air power on Leyte, or lose the war."

Suzuki was convinced that a coordinated effort between ground and airborne units would retake the Burauen airfields. The next paratrooper attack group was assembled on Luzon. The elite paratroopers arrived from Sasebo, Japan. The 3rd Raiding Regiment, was led by Major Tsuneharu Shirai, and the 4th Raiding Regiment was led by Major Chisaku Saida. Their aircraft remained on the island of Formosa to avoid U.S. observation until the start of the operation. The infantry units also prepared for the counterattack. The 16th and 26th Infantry Divisions moved east into the Central Mountain Range. It was planned that the airborne units would arrive first, then be aided by the infantry units in taking the airfields.

Apparently still unknown to the Japanese, U.S. fighter planes had stopped using the Burauen airfields. The only planes left functioning from the San Pablo airfield were the small L-4 Liaison planes, operating vital recon and supply missions to the 11th Airborne Division troops in the mountains. But even though no air strikes originated from Burauen, the area contained valuable service installations. The Dulag-Burauen road was lined with service units, mostly immobilized due to the rain-soaked conditions.

Previously, there had been at least three U.S. infantry battalions (3,000 men) guarding the area around the Burauen airfields. But as days passed since the paratrooper attack on November 27th, security there was relaxed. U.S. infantry were sent to the west coast of Leyte from all directions. A U.S. invasion force was on its way by sea to land near the west coast city of

Ormoc. Other U.S. infantry divisions, like the 11th Airborne moved west from Burauen to block possible escape routes through the mountains. The expediency of meeting MacArthur's timeline on Leyte left fewer than 200 combat-ready infantry troops at Burauen.

U.S. intelligence activities, known as "code-breaking," provided an extremely valuable advantage in the war against Japan. Intercepted messages were key to the Navy's victory at the Battle of Midway. Also, the U.S. shot down the plane of Japan's top Naval commander, Yamamoto, when intercepted messages revealed his whereabouts. The death of the "mastermind" of the Pearl Harbor attack was a great morale booster to the U.S. public.

As General Swing sat in his headquarters office at San Pablo Airfield, he received a coded message, called the "Alamo Report." American intelligence, "the code-breakers," had intercepted messages identifying a group of paratrooper transports that departed from Formosa to Luzon. The messages also indicated that the planes would carry paratroopers to Leyte to drop on the Burauen Airfields. The decoded message said that the Japanese were expected to attack with a brigade of paratroopers at 18:00 on December 6th. Swing apparently interpreted the report as reliable, although he doubted that the Japanese could launch a brigade-sized attack (3,000 to 5,000 men). He identified the available personnel in the area as two companies of the 187th Glider Infantry regiment at San Pablo, service troops consisting of the 127th Engineer Battalion, and the 511th Airborne Signal Company (less than 300 men). Pilots, mechanics, and supply personnel were scattered across the three airfields. The headquarters of the Fifth Army Air Corps, including General Ennis Whitehead's office, was set up in the village of Burauen, positioned near the Buri Airfield. These included the 287th Field Artillery Observation Battalion, with a detachment of 47 men, and a Marine Corps artillery unit, just north of Buri. A small number of 5th Air Corps support personnel, mechanics and pilots with limited combat training, occupied the airfields. The 44th was set up between the Buri and Bayug Airfields. Even with this intelligence, Swing did little but issue a warning for personnel in the area to be on alert. Apparently, he did not request reinforcements, instead determining to shift some of the personnel around first thing in the morning.

December 1, 1944

The Japanese 16th Infantry slowly made their way through the Central Mountains towards Burauen. General Shiro Makino had taken personal command of the division and would lead the counterattack. The Japanese faced challenges as they moved into position. The division originally numbered over 10,000 on A-Day, October 20th. By December 2nd Makino was left with only 500 men, most of them not healthy. This was indicative of the Japanese challenges in the Philippines and the War in the Pacific in general. They had difficulty supplying and sustaining their vastly expanded Pacific empire. It wasn't just bullets and bombs that took its toll. By the end of the battle for Leyte, out of 49,000 Japanese dead, 80% died from starvation or disease. A Japanese defensive strategy was to fight from fortified positions. The problem was that they could be surrounded by a superior U.S. force and cut off from their supply lines. The harsh environment of the Pacific islands and jungles would prove to be their downfall. The land offered limited sustenance, and many succumbed to starvation and disease that spread rapidly through their crowded bunkers and caves.

The Filipino guerrillas and villagers kept the 44th's officers informed of enemy movement in the area. They had recently observed elements of Japanese Infantry in the area. Mostly living off the land, the Japanese scoured villages and fields for food. These were likely members of the Japanese 16th Infantry.

Dad described a Filipino woman from a nearby village that came into the 44th's camp. "She was slight of build, but looked tough. She had a bolo knife attached to her belt. On the other side of her belt hung the head of a Japanese soldier. She requested to speak to an officer." Through an interpreter she told Dad and other officers of her exploits killing Japanese soldiers. Dad continued, "She said that she would sneak up from behind and dispatch them with one slash of her knife." Not wanting to question her story, the officers convinced her to let them take care of the head. She would return to the camp over a number of days with more prized possessions. Some in the 44th believed that she cut the heads off soldiers who were already dead. But her story may have been confirmed by an Associated Press (AP) newspaper article, dated October 26, 1944. It described a 38-year-old Filipino schoolteacher from Tacloban, Leyte, named Nieves Fernandez:

Captain Nieves Fernandez was a schoolteacher who became the only known Filipino female guerrilla leader. Working with guerrillas south of Tacloban, Miss Fernandez rounded up native men to resist the Japanese. She commanded 110 natives who killed more than 200 Japanese with knifes and shotguns made from sections of gas pipes. The Japanese offered 10.000 pesos for her head. She was wounded once. There is a bullet scar on her right forearm. In her battles, she was a master guerrilla fighter; an excellent crack shot, and hand-to-hand combatant. She helped liberate her island from the Japanese occupation, and the guerrillas also provided valuable intelligence during MacArthur's assault on the islands.

The 44th experienced increased sniper activity, particularly at night. In the mornings and evenings, Dad had his sergeants spray the tops of the coconut palms with machine gun fire to take out any snipers. At night, the 44th blacked out their tents. Under dim lighting for fear of snipers, the doctors operated in blacked out slit trenches. Staff were instructed to refrain from smoking in the open after dark. Many rigged up empty coffee cans to smoke in so they wouldn't be targeted. As the 44th had discovered, the Japanese didn't hesitate to shoot into the hospital, snipers reportedly even using the red crosses on the flags for target practice.

Japanese air attacks also increased starting on December 1st. At the 44th's camp, breakfast and dinner were interrupted for three consecutive days by Japanese planes. General Yamashita had ordered airstrikes on the Burauen area to prepare for the counterattack. The Japanese planes, based on the island of Cebu, strafed and dropped bombs around the airfields. Attacks were also aimed at U.S. anti-aircraft weapons protecting the area. As General Suzuki neared the time of the counterattack, he was about to be attacked himself as the Americans moved toward the west. But Suzuki had no intention of sitting idle while the U.S. surrounded him. Instead he seized the initiative and launched the attack to take back the prized Burauen airfields.

December 5, 1944

Suzuki's counterattack plan was unravelling from the start. Due to poor communications, he had problems coordinating the units for the planned attack. The weakened Japanese 16th Division was slowly moving south from Dagami to Buri. They camped in a canyon that limited their ability to communicate. The Japanese 26th Division, moving towards Burauen, got bogged down at Ormoc to the west. U.S. planes disrupted their supply convoy and they ran into part of the U.S. 7th Infantry.

Yamashita ordered the paratrooper attack to take place on December 5th, but bad weather forced a delay of one day. Suzuki reported that his infantry needed until December 7th to get into position. Yamashita agreed to the delay for the infantry's "Wa" operation, but the airborne attack on the Buri airfield was still set for 0630 on December 6th. General Makino, leading the 16th Division, was not notified of the change in plans. To protect the landing of the paratroopers, Yamashita dispatched additional fighter escorts. He also planned a diversionary attack at Limon, over 30 miles to the northwest of Burauen, to keep U.S. troops occupied through December 7th.

As the front lines of battle shifted to the west, it became apparent to the 44th's officers that infantry support had been prematurely removed from the area. They were now on the fringe of the combat zone and mostly unprotected. The increased Japanese sniper activity and multiple bombing and strafing runs raised concerns. The 44th's officers weighed the risks and discussed their options. What if Japanese infantry attacked in force? The 44th had limited arms to defend themselves, and over 90 percent of the staff had not been combat-trained. With over 200 wounded it would be very risky to evacuate. Evacuation by truck was not an option; the muddy, rain-soaked roads were now mostly impassable. A large hospital caravan would be an easy target for Japanese air and ground forces while trying to navigate back to the landing beaches. Evacuation by air was also risky, due to increased Japanese fighter activity in the area.

The main question was whether the Japanese would deliberately attack a hospital unit in blatant violation of the Geneva Convention? Experience across the Pacific indicated that the Japanese did not refrain from shooting medics and even firing on clearly-marked hospital ships. If attacked, what would be the likely fate of the doctors and hospital patients? Would they be killed or taken as prisoners? Thoughts of Solferino and the Bataan Death March came to mind. At Bataan, capture was a "slow death," as had been the experience of many U.S. POWs. If the 44th defended themselves, they could at least take charge of their own destiny.

To ease concerns and safeguard the hospital, Colonel Waddell requested infantry protection from Army headquarters. His repeated requests from December 1st through the 5th were denied. Headquarters staff did not believe that the Japanese had sufficient resources to attack at Burauen. Waddell then asked for authorization to obtain arms and ammunition for self-protection. The 8th Army Base Surgeon and the Chief of Staff refused on the grounds that "arming our medical officers and enlisted men would be in violation of the substance of the Geneva Convention." Being a hospital unit, they were expected to "do no harm."

The officers of the 44th debated their options. Most were convinced that the Japanese would not hesitate to attack the hospital, "Violation of the Geneva Convention? Hell, the Japs never agreed to sign it! Do we expect them to sing Christmas carols to us and deliver fruitcakes? Hell no!"

Given the choice between doing nothing or defending themselves, Colonel Waddell determined to arm the entire staff and setup a defensive perimeter. If the Japs chose not to cross into the hospital area, no one would know that they were armed and waiting. If they did cross, they would give 'em hell. According to the officers of the 44th, "We realized that we were on our own to defend the hospital." They would risk arming and fighting. They could still be overrun and killed, but at least they would go down fighting together. Officers, doctors, corpsmen, cooks, and supply men would together make a stand. Although some made grim jokes about being at the Alamo, possibly in reference to the "Alamo Report" received from Army Intelligence, the possibility facing a direct Japanese assault must have weighed heavy on their minds.

Colonel Waddell needed to obtain arms and ammunition for the entire staff. But without authorization, he had to quickly figure out how to do that. For Colonel Waddell, this could mean court-martial for disobeying an order. And what about the doctors and corpsmen? Was it worth the risk to arm them too? In basic training at Ft. Sill they were not even trained in how to shoot a rifle. The high-level command was adamant that, "They'll just need to know how to march." Combat was not something the doctors signed up for or expected to be part of their duty as a senior-level, attached medical unit.

For the doctors of the 44th, arming and fighting would be against their oath to "do no harm." But what about the harm that could come to themselves and their patients? If a senior medical unit and their 200 patients were massacred by the Japanese, what kind of backlash would the Army's leaders face? There would have been consequences for the lack of action to protect a vulnerable hospital. The press would have certainly exposed the blunder and there would have been outrage back home.

On the evening of December 5th, the Japanese 16th Infantry Division reached the foothills near Dagami, now within striking distance of the Buri Airfield. Only 300 men were left. Over 200 had been killed by U.S. artillery fire or by air attacks in the previous weeks. Morale for the unit was low. Besides their punishing journey, they were nearing starvation and racked with disease. They scrounged food off the land, mostly bananas and coconuts. Officers left the wounded behind to die. Relations between officers and men, never good in the Japanese Army, was at a low point. They rested for the night in a secluded canyon and prepared to attack. Their mission was to eliminate American air operations at Burauen, and they planned to strike at the Buri Airfield first.

December 6, 1944, Morning –
The Japanese Infantry Attack at Burauen

In the early hours of December 6th, General Makino's 16th Infantry Division reached the Buri airfield undetected. They expected to rendezvous with the paratroopers from Luzon and with the 26th Infantry arriving further south. Unbeknownst to them, the paratroopers had not yet arrived, and the 26th Infantry Division got delayed and fell several days behind schedule.

Members of the 287th Field Artillery Observation Battalion were the first to see the Japanese 16th Infantry Division emerge from the mountains to the west of the Buri Airfield. The Japanese crossed the main road and disappeared into the swamp that lined the airfield. They set up a machine gun in a shack 300 yards west of the roadway.

Led by a Filipino who was a Japanese sympathizer, the highly-experienced force moved to the airfield. Their first target was a bivouac area of the Fifth Army Air Force Bomber Command. They shot soldiers of a U.S. construction battalion while they were still sleeping. Some were bayonetted in their bunks before they could move. In hand-to-hand fighting the surprised U.S. troops fought with any weapon they could reach. Some grabbed rifles and pistols and started firing at the oncoming Japanese. An angry cook shot and killed five Japanese soldiers that broke into his mess hall's storage area. U.S. food supplies were a top priority target for the under-provisioned and hungry Japanese troops. The panicked firing caused the nearby 11th Airborne Division to complain of "friendly fire" from the excited Air Corps personnel.

The 44th was alerted by the shooting that emanated from the nearby airfield. They were shocked to see dozens of Air Corps service personnel fleeing from the Buri Airfield to the safety of the hospital, many without their clothing. Their tents had caught fire from incendiary charges used by the Japanese attackers. Some had their clothing burned off their bodies. They told of their tent mates being bayonetted in their sleep. The 44th provided first aid and distributed uniforms to those who needed them.

The 44th had very limited firepower to defend themselves if the Japanese moved towards the hospital perimeter. Dad's staff, responsible for hospital security, had a few Thompson submachine guns, some M-1 rifles, bayonets, grenades, and flares. Medical Administration (MAC) officers and sergeants also carried standard 1911 Colt .45s as sidearms. Dad assigned his sergeants and some enlisted men to guard the perimeter. But they knew they would be no match against a large Japanese assault. They waited anxiously as the 44th's Commander again sent out urgent requests for infantry assistance.

On the first day of the Japanese counterattack, the experienced 16th Division surprised and routed the American service personnel at the Buri Airfield. After clearing the airstrip, they positioned themselves into a wooded area just to the north. There they would wait to link up with the paratroopers who had not yet arrived.

December 6, 1944, Afternoon – Preparing to Defend the Hospital

Concerns for the safety of the staff and patients of the 44th was at a peak. It's not known how Colonel Waddell planned to obtain arms. The 44th numbered over 500 men, including MAC officers, doctors, corpsmen, and enlisted staff. Waddell determined that they would each need a rifle to repel a Japanese infantry attack. He ordered Captain Walter Hanna, MAC Supply Officer, and Captain Bill Hastings, MAC Registrar to gather a small group of enlisted men and proceed to White Beach in three ambulances. The supply depot at White Beach was approximately 10 miles to the east. They would have to navigate on the muddy roads and face swollen rivers and treacherous bridges. They also ran the risk of running into a Japanese patrol or an airstrike.

At the hospital's perimeter, Dad and his security detail anxiously awaited the return of the ambulances. What if they didn't make it back in time? What if they didn't obtain the rifles? How long could the limited group hold out against a Japanese infantry assault? Rifle and machine gun fire could be heard at the Buri Airfield just a few hundred yards away. The men nervously watched for any enemy movement through the thick brush and tall grass.

After a slow and tedious trip, the 44th's ambulances reached the beach. Then somehow, without official authorization, 400 M1 carbine rifles were obtained from the ordinance unit. Captain Hastings recalls the acquisition, "We took three ambulances to haul all the guns and ammunition. We took several enlisted men to assist. The 8th Army Base Surgeon did not support us in this matter. I give Hanna the credit for obtaining the rifles." Sometimes it matters "who you know" or maybe even more, what you have to bargain with. Did the 44th's commanders have a connection at the ordinance depot? Was there a trade for a stash of whiskey and gin, accumulated from a year in Australia? Did they get authorization under the guise of transporting to the 11th Airborne Division stationed nearby? No one has ever stated how the arms were obtained and no official record of requisition has been found.

That afternoon, to the relief of all, the ambulances returned from White Beach. The back doors of the three ambulances opened, but there were no wounded inside to unload. Instead, enlisted personnel started unloading the crates of rifles and ammunition. Now the rifles had to be distributed and the men quickly trained in how to use them. Anyone experienced with firearms was called upon to assist.

Captain Ray LaFauci recalls the events, "I was asked by Ed Odrowski to start instructing the EM (enlisted men) in the use of the newly arrived M1 rifles. My limited 'expertise' derived from my exposure at OCS (Officer Candidate School) to a single afternoon on the firing range. I only completed a couple of sessions with the EM and never got beyond field stripping."

The newly acquired M1s were covered in cosmoline, a.k.a. "shipping oil," a rust preventive. Cosmoline is a brown-colored, wax-like substance that has a petroleum-like odor. It was a real "pain in the ass" to clean off the rifles. As the rifles were issued, the men were instructed in "field stripping," i.e., disassembling the rifle so it could be cleaned. The cosmoline residue had to be removed, else there was a risk of fouling the operation of the rifle. Having a jammed rifle was not a good situation if being charged by Japanese soldiers with bayonets.

The M1 carbine is a lightweight, compact rifle that basically fires a long pistol cartridge (the .30 carbine). The M1 Garand is the larger, general battle rifle for the infantry, firing full power cartridges (.30-06) and capable of engaging enemies effectively at 600 or more yards. M1 carbines were usually less accurate and less powerful than the longer barrelled rifles of the infantry, due to a shorter sight plane and lower velocity of bullets fired from the shortened barrel. But the advantage of a lighter-weight rifle, 4 pounds vs. 9 pounds for the Garand made a big difference in harsh conditions as on Leyte. During World War II and since, most infantry combat has been fought within 150 yards. The bigger problem with the M1 Carbine versus the Garand was reliability. The Garand is a very reliable rifle, particularly considering that it's semi-automatic. The same cannot be said for the M1 Carbine, which was known to jam.

Richard C. Janes was a supply sergeant with the 44th. As part of the medical unit, he had not received combat training. But being a long-time deer hunter from Portola, CA, he was proficient with a rifle. After the rifles were distributed, he lined up some of the doctors and provided basic arms training. He and Supply Corporal Manuel S. Mathews then took their positions at the hospital perimeter.

The 44th hastily implemented their defense plan. With the aid of 100 Filipino laborers a defensive perimeter was set up around the hospital area. Individual as well as large fox holes were dug. The 44th planned to man the foxholes around the clock with the newly acquired M1s.

Dr. Chet Gjertson, the 44th's dentist, recalls setting up the perimeter defense, "We had first set up a walking defense where we'd walk around the perimeter on patrols looking for Jap activity. This proved to be a bad idea as we started drawing fire from snipers. So instead, with the help of Filipino laborers, we dug large foxholes all around the perimeter. The plan was to put two officers in each corner, 30 to 40 feet apart, with armed enlisted men in between. The EMs were told to stay put and fire on anything that moved."

Dr. Sims recalled his time in the foxhole, "I was commanding the southwest corner of our perimeter. During the day I would memorize every rock, stump and clump of brush. I'll still swear that they got up and walked around."

Dad recounted a procedure that he used to protect the perimeter: "In the evening I had my sergeants string trip wires from trees about 50 yards out. They took a hand grenade and rigged the pin to the wire. They then nailed the wire into a tree at ankle height. If a Japanese soldier advanced in the dark and tripped on the wire, the grenade's pin would be pulled and it would drop and explode." He then readied the sergeants with flare guns to light up the area in the direction of a grenade explosion. The flares would expose an advancing enemy unit and provide visibility to open fire upon them with machine guns and rifles.

Word reached XXIV Corps Headquarters of the morning attack. General Swing, recalling the intelligence reports he had received, mobilized some infantry units to move to the Buri Airfield. When they arrived at the airfield, they had to calm down the excited Air Corps personnel that remained. Army HQ was still unaware that a major Japanese counterattack was underway. But as the number of Japanese reported in the area rose, it was determined that this was a considerable force. Swing formed a battalion of 180 infantrymen. They formed lines of skirmishers positioned six feet apart and attacked the Japanese at the Buri Airfield. Moving to the northeast, they combed through dense jungle growth. Soon they were in firefights at point blank range in the thick, tall grass. Japanese seen in the rice paddies were more visible and easily killed by riflemen. After a couple of hours, 85 Japanese were reported dead, with only two U.S. casualties.

Several hundred more Japanese were reported coming from the hills to the west. U.S. artillery fire disrupted their movement, but had to stop when shells landed too close to U.S. positions. Swing mobilized the 187th Glider Infantry from the landing beach. By evening, the 1st Battalion, 187th Glider Infantry had cleared the Buri Airfield of all but a few pockets of Japanese. Under constant sniper fire, they continued to hold at the airfield.

December 6, 1944, Evening – The Second Paratrooper Drop

The second wave of the Japanese counterattack, this time by air, was about to hit the Burauen airfields. The airborne assault by the men of the Japanese 2nd Parachute brigade was planned for the evening of Dec. 6th. Some of the paratroopers would board Type 100 heavy bombers. Their mission was to land on the Burauen airfields and destroy American aircraft and supply dumps with demolition charges. Other paratroopers would jump from Type 100 Transport planes to engage security forces and destroy antiaircraft defenses. Of the total airborne force, approximately 250 paratroopers were assigned to the Buri Airfield, 72 to the Bayug Airfield, and 36 to the San Pablo Airfield.

The jump would be led in three waves by Lieutenant Colonel Tsunhiro Shirai. They would depart from Clark Field, north of Manila on Luzon. The paratroopers were armed with automatic weapons, land mines, and dynamite loaded in satchels. Shirai carried a flag inscribed with "Exert your most for your country," signed and given to him by the commander of air operations in the Philippines, Lieutenant General Kyoji Tominaga. Paratroopers carried neatly folded cramming notes in their pockets with English phrases such as "go to hell, beast." They were also presented with special bottles of sake with labels stating, "not to drink until ready to jump." Sake was issued to boost morale, particularly on suicide missions. "Exerting the most for their country" can be interpreted as "succeed or die trying." A cameraman from the Nippon News documented the mission planning, preparations, and departure. In Japan, this was shown to moviegoers in a newsreel called "Leyte Paratrooper Attack." The Japanese paratroopers were considered an elite fighting unit. The men were well-educated, highly-trained, and equipped with advanced weaponry. They were even better-dressed than most units. Many wore ceremonial sashes and a hachimaki, the headband worn by samurai, and later the kamikaze pilots. The Japanese media lauded them as heroic "Shinpei," translated as "soldiers dispatched by the gods."

THE BATTLE OF BUFFALO WALLOW

At 1540 Hours, thirty-five transports and four heavy bombers lifted off from Clark Field on Luzon bound for Leyte. The transports flew southeast over Bacolod Airfield on Negros Island where they met up with light bombers and fighters, then turned eastward towards Leyte. They crossed over the Central Mountain Range at the start of the evening. To avoid detection, they flew just above the mountain tops with the setting sun behind them. Members of the U.S. 11th Airborne, positioned high in the mountains were almost at eye-level with the planes. They watched in desperation as the planes flew past. Their radio operators tried to alert the U.S. positions at Burauen but were unsuccessful. An early warning of the raid would have saved many U.S. lives.

The Japanese transport planes, similar in appearance to U.S. DC-3s, arrived over the airfields at twilight on December 6th. The skies were relatively clear. An impressive sight, the 35 transports formed a "V of Vs" as they approached the drop zones. Flying slow and at an altitude of only 700 feet, the paratroopers prepared to drop. Major Shirai and about sixty of his 3rd Raiding Division jumped over the Buri Airfield. American anti-aircraft positions, finally seeing the planes, responded with a heavy barrage. Due to confusion caused by the firing, many pilots missed Buri and dropped most of the paratroopers over the San Pablo Airfield. At San Pablo, one plane's anchor line failed to pull the rip cords causing all of its paratroopers to drop to their death. Their equipment may have been damaged by the artillery fire. Some paratroopers fell 200 yards from the San Pablo runway and got tangled up in the tall trees. All transports that headed towards the operational airbases at Dulag and Tacloban were shot down with the paratroopers still on board. Of the 35 original planes that left Luzon, only 17 would make it back. The ones that returned were heavily damaged.

As the 44th finished dinner and clean-up that evening, a flight of planes was heard and seen approaching the airfields. Many again thought that they were U.S. C-47s coming in to drop off wounded. They watched transfixed as antiaircraft guns opened fire on the planes. Japanese fighter escorts arrived first, strafing the airfields and surrounding area. Bombs were exploding around the airfields and the antiaircraft gun positions. Japanese planes hit by flak were disintegrating overhead. The 44th's staff ran for weapons and took cover in their recently dug foxholes, dodging fiery debris as it fell from the sky. Looking up, they saw parachutes dropping from the sky. Over 250 jumpers from the Japanese 2nd Raiding Brigade had arrived. Groups that landed first set off smokescreens to conceal those who jumped after them.

Dad said that he was out near the rice paddies a hundred yards or so from the hospital when the planes came in. Sixteen Japanese paratroopers drifted to the west of the Buri Airfield, missing their target. Dad described watching this group of paratroopers land between him and the hospital perimeter. He claimed that he was only armed with a golf club and a musette bag full of golf balls that had recently replaced his grenades. He watched as the planes continued to the other nearby airfields dropping more paratroopers in the distance. Anti-aircraft artillery was firing and hitting some of the planes. The paratroopers in front of him, recognizing that they were off-track and alone, picked up their gear and moved to rendezvous with the others that dropped on the Buri Airstrip. Relieved that the wayward paratroopers had moved on, Dad cautiously worked his way back to the hospital perimeter.

Chet Gjertson, the 44th's dentist, described the Japanese paratrooper drop,

I recall seeing the Jap transport planes fly directly over the hospital and descend on the Buri Airfield a few hundred yards away. I thought that they were U.S. C-47s coming in. Tokyo Rose, on a recent radio broadcast, warned that if we didn't leave Leyte soon, that we'd get gassed off the island. When we saw smoke from smokescreens and burning fuel, someone cried out "gas" or "Japs." Everyone scrambled in panic looking for their gas masks. Some took cover in their tents as gas wouldn't come in through the tent canvas. The chaplain, running for a foxhole and without a mask, asked if I had an extra one. Fortunately, it was just smoke, not poison gas.

Other U.S. troops, who also thought that the planes were theirs, saw men standing in the open doorways. They recognized that they were not American. The U.S. soldiers rushed to their tents to grab weapons. Officers fired .45 sidearms at the paratroopers as they slowly drifted to the ground. George Mendenhall, with a Marine observation unit stationed at the Buri Airfield recalled the drop, "I was there when the paratroopers were dropped and shot at them on their way down with my .50 caliber machine gun. Our unit, 18 of us, held the airfield until an Army unit arrived."

Henry J. Miller, Jr. recalls,

At first it sounded like a swarm of bees in the distance. Then it became clear. No one could mistake the drone of a formation of troop carrier aircraft. Someone outside shouted "Transports! Japs! Paratroopers!" The division staff dashed out of the mess tent looking skyward. By now a dozen parachutes had opened above us and everyone began firing at them. I even emptied two clips from my .45 at the nearest parachutists.

Frank Widay, with the 892nd Chemical Company, recounted shooting at the paratroopers dropping on San Pablo, "This was the only time (in the War) that I fired my personal weapon."

Once on the ground, the paratroopers used a system of bells, whistles, and horns to assemble. Several spoke in loud voices, repeating in English, "Everything is resistless, surrender, surrender!" or "Hello, where are your machine guns?" Confusion was apparent among the attackers. Many were killed before they could take up fighting positions. Some who may have drunk too much of their special sake, moved randomly about screaming "banzai." Once they were organized, they moved towards their objective, the airfields. Some set to work blowing up gasoline dumps at

San Pablo and Buri airfields. At Buri they set fire to a number of the small liaison planes of the 11th Airborne Division, then moved to a bivouac area and destroyed it. At San Pablo Airfield the Japanese destroyed more small planes, a jeep, and several tents. They threw U.S. ammunition into the fires to destroy it as well. At Bayug, some paratroopers were shot as they landed. Others managed to reach a line of liaison aircraft and blew them up with grenades. They also set fire to fuel and supply dumps. The 44th General Hospital was in between the burning planes and fuel at Buri and Bayug. Thick black smoke filled the air and obscured the remaining daylight.

Private Mort Ammerman of the 188th glider regiment also thought the planes were U.S. C-47s, making a night jump. He went to sleep in the pouring rain to be woken up by the sound of gunfire. A pain in his leg indicated that he had been hit. He couldn't find his rifle. Leaving a dying comrade, he hobbled away to take cover with only a trench knife. Later, in the darkness, he could hear the Japanese taunting them in poor English.

The startled 187th Glider Division and the Air Corps service troops caught in the middle of this attack fought for their lives. At the 187th's Battalion Aid Station, the Battalion Surgeon, Captain Hans Cohn was working on a wounded soldier when the Japanese shot a plasma bottle out of his hand. He then had the man lowered into a foxhole and completed his work.

U.S. 11th Airborne infantry dug into their defensive positions. Being paratroopers themselves, the shock value of an airborne attack was lost on them. At one end of San Pablo airfield, Lt. David Carnahan set up a machine gun post with 40 U.S. personnel. They observed a column of men marching along the airstrip singing "Sweet Adeline." Confused as to why U.S. troops would be so casual in this tense situation, they waited until they got closer. When the singing group got within 20 yards, Carnahan recognized that they were Japanese. A Japanese officer, in a heavily-accented voice, called out "is this the machine gun at the west end of the strip?" Carnahan answered "yes sir," and promptly opened fire on the imposters. Many of the Japanese paratroopers were wearing U.S. Army uniforms under their jump smocks.

General Swing was finishing a fried chicken dinner in his office at San Pablo when the paratroopers dropped on his 11th Airborne headquarters building. Swing was convinced that this was part of the Japanese counterattack indicated in the "Alamo Report" several days before.

The Japanese airborne raid was an initial success. It turned out to be one of the few large-scale Japanese airborne assaults of the War. Within an hour, the paratroopers succeeded in burning U.S. liaison planes, destroying fuel dumps, and killing many administrative and maintenance personnel. The American support troops were not prepared for such an assault. Japanese paratroopers held the Buri airstrip by dawn on December 7th. Some Japanese paratroopers also moved to the north and linked up with the remaining members of the 16th Infantry Division. It was expected that they would combine and attack in full-strength after dark.

At the same time, General Krueger's plan for the amphibious attack on the west coast of Leyte was in-progress. A convoy carrying the 77th Infantry Division was in route to the beaches at Ormoc. As night fell on December 6th, Krueger received some unexpected news. He learned that a large number of Japanese planes were reported over central Leyte and that close to 50 planes had been shot down. The most disturbing news was that a fleet of enemy transports was dropping hundreds of paratroopers on the airstrips near Burauen. Attacks were also reported at the Dulag and Tacloban airfields. General Ennis Whitehead, commander of the Fifth Army Air Force, had reported that his headquarters was under heavy ground attack. He also reported paratrooper drops in the area of the Buri and San Pablo airfields. As it turned out, General Suzuki had struck first.

Private First-Class Donald O. Dencker, was a member of Mortar Section, Company L, 382nd Infantry Regiment, 96th Infantry Division, that landed on Leyte. Pfc. Dencker had just finished dinner on December 6th at 7 p.m. when new orders arrived. He was to be combat-equipped and ready to move out in 20 minutes. Climbing into a two-and-a-half-ton truck, Dencker and his unit proceeded down a muddy and bumpy road in the dark. The truck moved slowly, 15 to 20 miles per hour, due to the poor roads. It was decided that it would be safer and quicker to walk, so they unloaded and moved on foot down the road. In the distance they could hear sporadic gunfire. An occasional flare or flame would light up the night sky as they approached the airfields.

After walking another 30 minutes, a large group of tents were seen in the darkness beside the road. They arrived at the 116th Station Hospital (SH) and prepared to guard the facility and its staff. They had not yet taken in casualties. The 116th SH was north of the town of Buri, approximately two miles from the Buri airfield. The hospital covered a large area that was cleared in the jungle. Company L spread out around the perimeter. They heard the chaotic sounds of battle to the south. Frequent cracks of rifle fire, occasional burst of machine gun fire, intermittent explosions, and the racing of truck and jeep engines, presumably stuck in the mud. Members of the Company L rifle platoons started teaching hospital staff how to use the M1 carbines. Initially an unarmed medical unit, the hospital was now being provided with small arms and training for self-defense.

At the Buri Airfield, John Tilley of the 431st Fighter Squadron recalls, "Pilots spent two nights in a row on the ground until the Army cleaned out the Japanese paratroopers. We did not get any sleep for two days (sitting) in a slit trench with our M1 carbines. The pilots were so nervous that if a rabbit had moved it would have been blown to hell."

The Japanese infantry had in fact moved from the west coast of Leyte into the mountains to the west of Burauen, in front of the 44th. The Japanese paratroopers that dropped on the Buri airfield were less than a quarter-mile behind. The enemy was not only in front but also behind the hospital. It was a tense night for the 44th personnel who manned the perimeter waiting for a possible attack. It had at least been a relatively dry night. The foxholes were not filling up with rain water as in previous nights before. With over 400 rifles on the perimeter, this time the 44th was well-armed and ready.

Doctors of the 44th General Hospital.

Dad and Mom in
Lompoc, CA.

Dad in basic training at Camp
Grant, IL with World War I era
uniform and gas mask.

Walter Teague, 44th GH Mess Officer, with wounded Japanese prisoners.

Colonel Waddell and Colonel Bechtold,
commanding officers of the 44th GH.

Mom and son Edward "Eddie", born while Dad was at Ft. Sill, OK in 1943.

Frank Sadecki, Mom's older brother, Navy Radar Operator on the Destroyer U.S.S. Little, hit and sunk by Japanese kamikazes at Okinawa.

Chester Sadecki, Mom's younger
brother in the Navy.

Dad at camp in Australia, before the
War got a lot more interesting.

Kiki native leaders on New Guinea. Dad is on the back row farthest to the right. Colonel Weston and doctors to his left. The Kikis in the middle are elder women.

Dutch missionaries rescued at New Guinea. Dad is on the second row, second from the right.

Filipino version of "4-wheel drive", a carabao hauling supplies through the mud at Burauen. Dad is on the left.

The 44th General Hospital on Burauen. Leyte's Central
Mountain Range is to the left, the Buri Airfield behind
them and to the right.

Sgt. Zeumer on the Daguitan River.

Dad at Burauen,
looking particularly
thin.

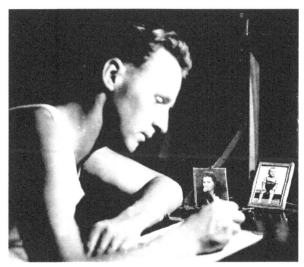

Dad writing letter home from inside his tent on Leyte.

Chet Gjertson on the left and other 44th GH doctors
proudly posing with their M1 rifles after the Battle of
Buffalo Wallow.

Japanese paratroop leader Lieutenant Colonel Tsunhiro Shirai who led the December 6th, 1944 airborne attack on the Buri Airfield near Burauen.

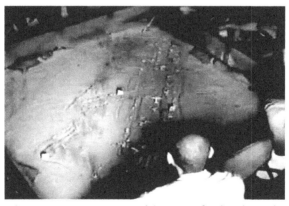

Japanese paratroopers studying maps for the planned attack on the Burauen airfields..

Japanese prisoners in hospital stockade at
Burauen. Dad stands behind them supervising
some type of activity (doing an art project?).

Filipino children, likely orphaned, standing by the river at
Burauen. Sgt. Zeumer is behind them.

The look of defiance in both father and son, both
armed with bolo knives. The bolo the father is
holding is similar to the one that Dad brought home.

The smiles of the children
return with the liberation
from the Japanese.

Very young Filipino girls carrying
buckets.

Dad and Filipino girls from a
festival parade.

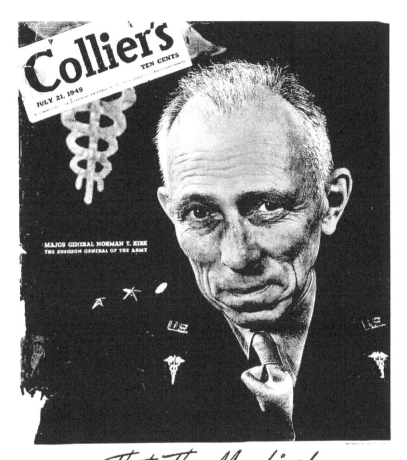

Surgeon General of the Army Norman T. Kirk on the cover of
Collier's Magazine, July 21st, 1945. In the article he provides a
glowing account of the 44th's service on Leyte.

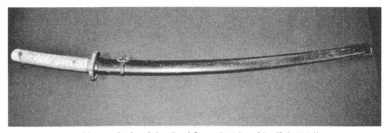

Japanese sword brought back by Dad from Battle of Buffalo Wallow on Leyte.

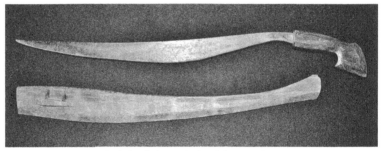

Filipino bolo knife brought back by Dad from Leyte.

Mom and Dad at Dad's retirement party in 1979. Dad had worked at
Colgate-Palmolive for over 30 years.

Kiowa painting of The Buffalo Wallow Fight, part of the Red River War,
September 10th, 1874.

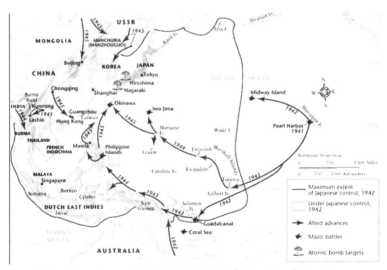

Major campaigns of the World War II Pacific Theatre of Operations.

Map of the Leyte Invasion October 20, 1944.

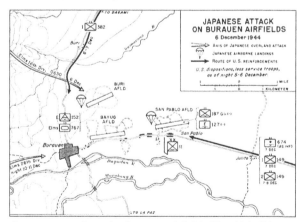

Japanese attack on the Burauen airfields. The 44th GH was positioned on the north-south Dagami-Burauen Road, in between the Buri and Bayug Airfields. Note the path of the Japanese 26th Infantry Division on December 10th. It brought them into the 44th's perimeter.

December 7, 1944 – Fighting at the Airfields

At daylight on December 7th, members of the 11th Airborne and other units prepared to attack the Japanese paratroopers at the San Pablo airfield. Army engineers were lined up abreast along with the infantry and a cadre of service personnel, including some from the 44th. It was reminiscent of a Civil War charge. The group moved in unison across the flat ground towards the enemy. To one Midwesterner of the 44th, it reminded him of his pheasant hunting days in Wisconsin. Back home, large hunting parties would line up and cover a large section of a farmer's field to flush out the birds. As birds were flushed, men shot at the ones that flew within their range. Similarly, when a Japanese soldier was uncovered, he was quickly covered by rifle fire. As the Japanese retreated from the San Pablo Airfield, the American units set up a defensive perimeter.

At the Buri Airfield there were few U.S. defenders, as it had been largely abandoned after the Japanese 16th Infantry attack the previous morning. The Japanese paratroopers seized arms and ammunition left by fleeing Air Corps and Service Troops after they arrived later that evening. The Japanese dug in to hold the field, waiting for other ground troops to arrive.

As General Kreuger learned of the size of the Japanese attacks at Burauen, he requested additional troops from MacArthur. Members of the 38th Infantry Division, originally staging on Leyte for the invasion of Luzon, were sent to the area. The 674th Parachute Field Artillery Battalion was ordered to leave its big guns on the beach and proceed to the airfields to fight as infantry. Upon arrival at Burauen the reinforcements were assigned to General Swing, 11th Airborne Division. He ordered them to take back the Buri Airstrip by nightfall. Navigating across the swamp slowed their advance and stronger than expected Japanese opposition forced a stalemate by late evening. The newly arrived U.S. battalion dug in for the night opposite the Japanese.

Throughout the day of December 7th, the 44th continued to hear the sounds of rifle and machine gun fire, plus an occasional mortar shell explosion, from the Buri Airfield. Captain Willard Arnquist, a doctor with the 44th, recalled the activity,

> *Our artillery was firing from some distance behind us. The shells*
> *sounded like freight trains passing over us. The shells exploded on*
> *the side of the mountain where Japanese infantry were digging in.*
> *We saw a Japanese Zero being chased by a pair of our P-38s. The*
> *Zero took fire from the P-38s then crashed in flames on the side of*
> *a mountain.*

On December 7th, the U.S. Army 77th Infantry division landed at Ormoc on the west coast of Leyte. The Japanese would soon lose their ability to reinforce Leyte from the west. A second wave of paratroopers took off from Luzon, but bad weather forced them to abort the mission. For the Japanese, their odds of saving Leyte were getting slimmer.

December 8, 1944

On December 8, 1944, the 1st Battalion of the 187th Glider Infantry arrived to take over clearing the airfield. Three U.S. battalions now linked up at the Buri Airstrip. But the Japanese were not yet defeated. Overnight, they brought up additional machine guns and placed them opposite the Americans. When Company A, 382nd Infantry started to move towards the Japanese positions, they were immediately pinned down by machine gun fire. A barrage of mortar fire was returned by the Americans that knocked out a number of the Japanese guns.

American infantrymen persisted in driving the Japanese from the airstrip. Ova Arthur Kelley was a 30-year-old private in the 382nd Infantry Regiment, 96th Infantry Division, Company A Weapons Platoon. Pvt. Kelley was from Norwood, Missouri, a small town nestled deep in the Missouri Ozarks. According to his platoon leader, Pvt. Kelley "just got mad" and launched a one-man attack on the Japanese. He charged out of his foxhole with an armful of grenades, throwing them at the stunned

Japanese. Five of the enemy were killed and the rest retreated. He then picked up an M-1 rifle and emptied it, killing three Japanese as they fled. Dropping the empty rifle, he picked up another M-1 and continued firing, killing three more. Inspired by his actions, the rest of the platoon advanced and finished off the 36 enemy soldiers that had pinned them down earlier. Still not finished, Pvt. Kelley led the platoon in an attack across the airstrip. As the company completed its objective, Pvt. Kelley was mortally wounded by a sniper. He was taken to the 44th General Hospital and died two days later. For his gallantry and leadership on December 8, 1944, Pvt. Kelley was awarded a posthumous Medal of Honor. The bridge carrying State Highway E over State Route 60 in Norwood, Missouri, was designated the "Pvt. Ova A. Kelley Medal of Honor Memorial Bridge" in his honor.

Officers Walter Teague and Jim Bingham of the 44th walked over to the mud-bound airstrip only to find that the clean-up was not quite over. There were both U.S. and Japanese dead and wounded there. Some GIs were in tears over the loss of their companions. Walter recalled the scene,

Jim commandeered a tractor-like vehicle to carry a badly wounded GI back to the hospital. It appeared that he had been bayonetted several times. It was not known whether he survived. Two wounded Japanese were also brought back to the hospital for treatment. They survived and were later turned over to Army Intelligence officers for interrogation.

December 9, 1944

In the morning hours of December 9th, the 1st Battalion, 149th Infantry attacked the Japanese with all three companies. They crossed the Buri Airfield under enemy fire. The Japanese guns were located on the high ground to the north of the airstrip. The Japanese were cleared from the airfield but the battalion was forced to the southern end of the airstrip to avoid machine gun fire from the ridge. Fifty enemy dead had been counted. As the battalion's patrols fanned out over the area, they left behind a small number of mortar men and support staff. At nightfall, this group was attacked by over 150 Japanese. They held off the attack, killing 50 while losing seven.

By the evening of Dec. 9th, two weeks after the first Japanese paratrooper assault on November 27th, General Swing had seemingly brought Suzuki's surprise counterattack under control. The infantry believing that they had now "cleared" the Japanese from the Burauen area, sent a Signal Corps platoon to guard the 44th. During the afternoon of December 9th, the platoon arrived at the 44th. They were fed and given an area to store their packs. They set up three .50-caliber machine guns in a ravine at the lower left corner along the hospital's western perimeter. Even though they were combat-trained and armed with three .50-caliber machine guns, they would be no match for a large Japanese assault.

On the evening of December 9th, Filipino informants again warned the 44th of Japanese activity in the small villages to the west of Burauen, along the narrow road that snaked through the jungles of the Central Mountains. The night was another tense one for the 44th. The men waited and watched from their foxholes. Some took turns sleeping while others watched for enemy movement around the buffalo wallow to the east. As they would find out the next day, the Japanese were not done yet.

December 10, 1944 – The Battle of Buffalo Wallow

At dawn on December 10th, the members of the 44th were waking up after another restless night in their foxholes. There would be no church services on this Sunday morning. Instead, the chaplains led small prayer groups. Some prayed silently. The silence was broken when the U.S. infantry unleashed a half-hour artillery barrage that rocked the mountains to the west. Following the artillery fire, the U.S. 1st Battalion, 149th Infantry, gathered just north of the hospital. They attacked the Japanese at the Buri Airfield with three assault companies. By late afternoon they had cleared the area of individual Japanese riflemen and destroyed small pockets of resistance. The battalion then dug-into a defensive perimeter to the north of the airstrip to face the remaining Japanese of the 16th Infantry who were holed-up in a wooded area.

The 44th's staff continued to care for the wounded during the day. Some of the officers resumed training doctors and corpsmen in handling the M1 carbines. Everyone still felt uneasy about the possibility of another Japanese attack. There were new reports of Japanese activity to the west. Also, the lack of infantry support in the area was apparent. Besides the Signal Corps platoon ahead of them, other infantry units had mostly moved to the north and west. If the Japanese attacked, the 44th expected that they would follow their pattern of attacking at night.

The 44th prepared to defend itself and its patients. As evening set in, Colonel Weston assembled a formation of doctors and enlisted men and asked, "Do you guys want to stay alive?" A chorus of "yesses" followed. They were instructed to proceed to their assigned positions. Armed with their carbines they waited in their foxholes for night to set in. To make matters worse, a steady rain started falling and the foxholes were filling with water. Standing water had to be intermittently bailed out. Wooden boards and bamboo poles were used to firm up the muddy ground beneath them. Poor visibility on this dark, dismal, rainy evening, made the scene in front of them seem even more threatening. If the Japanese infiltrated the camp, the untrained defenders may be faced with close quarters combat.

Jackman Pyre of the 44th recalls the early evening of December 10th,

> On December 10th at dusk, several planes flew towards the camp. Some thought that they were our planes as they came in low to approach the airfields. I noticed blue lights flashing. Someone called out, "Hell, those are Jap guns and they're shooting at us! Take cover." I jumped into what I thought was a foxhole. It turned out to be a garbage dump and I was soon covered in ants.

The Japanese 26th Infantry had finally arrived. They had worked their way slowly from the west on a steep, narrow trail through the Central

Mountains. Having endured harassment from the U.S. 7th Infantry and artillery from the 11th Airborne, only 250 men remained. They advanced through the outskirts of Burauen, alerting some of the Filipino villagers. Worn out and hungry, they looted village homes for provisions. Although weakened significantly they were dedicated to their original mission, to attack the Bayug Airfield. The paratroopers they expected to meet up with had landed four days before and were now mostly dead. They also expected to rendezvous with the 16th Infantry whom they had fought with before. But the 16th had also been decimated and was stuck fighting U.S. infantry to the north at the Buri Airfield. As the sun was setting, the 26th Division observed the Japanese air strike in progress. It gave them renewed hope that the counterattack was proceeding as planned. They ate and rested, waiting for darkness. Some drank liquor they had taken from village storerooms. Between them and the Bayug Airfield was the "buffalo wallow" and the perimeter of the 44th.

At 9:30 p.m. the Japanese 26th Infantry started their advance. The remnants of this once formidable jungle fighting force crossed the Dagami-Burauen Road and walked in the direction of the Signal Corps machine gun post. Startled by the advance, the U.S. platoon's lieutenant gave the order to start firing. They sprayed the area ahead of them with their machine guns. The Japanese took cover behind a grove of trees and returned fire. Some moved in the darkness, under the cover of their own guns, to approach the Signal Corp's position. The machine gun crew held off the advance until a grenade landed in their pit. It exploded, mortally wounding their lieutenant. The Japanese, screaming "banzai," charged the machine gun crew with their bayonetted rifles.

At 10 p.m., rifle and machine gun fire hit the administration buildings of the Fifth Army Air Force. Major General Ennis C. Whitehead was at his desk when a bullet came through his wall. After ducking for cover, he ordered his aide to find out who was responsible for the indiscriminate shooting. The staff officer contacted Lt. Colonel Paul V. Kaessner of the 8th Aviation Signal Battalion by telephone. The following conversation is reported in the Army's official history:

"Colonel," the aide said sternly, "you've got to stop that promiscuous firing down there immediately!"

"Like to, sir," answered the Colonel, "but the Japs…"

"Japs" shouted the staff officer, "that can't be Japs. That fire is coming from our fifties (.50-caliber machine guns)."

"That's right…and the Japs are doing the shooting!"

"Where in the hell did the Japs get our machine guns?"

"How in the hell should I know, sir?"

"The bullets are coming right through the General's quarters."

"Tell the General to get down on the floor. Incidentally, that yelling you hear is a banzai raid on our mess hall."

Dad described the start of the attack on the 44th's perimeter, around 10 p.m. that night,

> We waited in our foxholes, anticipating that the Japanese would attack after dark. We heard gunfire ahead of us where the Signal Corps platoon was set up. There were explosions and more firing. It got quiet for a few minutes. Then directly in front of us, we heard the Japs taunting us. Some were evidently drunk on sake, calling on us to surrender. We waited to fire, not wanting to give up our positions.

Jackman Pyre continues his story of the night, "Around 10 p.m. I was awakened by gunfire and all hell breaking loose. Japanese soldiers were calling to each other and to us. I crawled out of the dump through rain and mud to my assigned foxhole. It was on the left corner of the perimeter facing the buffalo wallow (swamp). The Signal Corps' machine gun post was to my left. On the way to the foxhole the rain made me shiver with cold and fear. But I was temporarily warmed by pissing in my pants."

Dad's perimeter group watched anxiously to see if the Japanese would give up their position. He recalled, "A grenade dropped from one of the trip wires and blew up. One of my sergeants shot a flare in the direction of the blast. The field looked bright as day as the flare slowly fell. We covered the area with machine gun and rifle fire."

Jackman Pyre recounted that fearful moment in the foxhole,

I stood watch with an enlisted man. Suddenly the flash of a star shell (flare) illuminated the area ahead. To my horror I saw a dozen mud-covered bodies crawling towards me. In my greatest moment of fear, I managed to get out a "piggly," the challenge password for the night (as the Japanese have difficulty pronouncing an English "L" sound). To my great relief I was answered by a barrage of "wigglys" from the retreating Signal Corps platoon. Lucky for them that they weren't shot, as we were told to fire upon anything that moved. They had been overrun in a Japanese banzai charge. Their lieutenant had been killed in the ambush. They abandoned their machine guns and retreated to the relative safety of our camp.

Captain Ray LaFauci also recalled the Signal Corps retreat,

A tense moment occurred when soldier after soldier came crawling up to our position, calling out "piggly, wiggly" and "sorry 44th, we can't hold our position." The fact that they were combat-trained and well-armed seemed to be a grievous breach of Army discipline. The next day found their machine guns in place and fully operational.

Pyre described some of the panic that ensued, "Some Japanese officers were calling out instructions to their men. They shouted 'banzai' talk to frighten us, which it did. An Italian private from Brooklyn, who worked in the mess unit, was positioned in a foxhole out in the wallow a bit further from the rest of us. With the Japanese approaching, he hit the panic button and started screaming hysterically. Colonel Weston asked for a volunteer to help retrieve him. The Colonel and volunteer went out in the darkness, pulled him out of the foxhole, and brought him back to safety. The episode left him a psychic casualty and he was subsequently sent home early."

Captain Chet Gjertson recalls the battle,

THE BATTLE OF BUFFALO WALLOW

I was positioned on the north side of the hospital perimeter facing the area where the paratroopers had landed. I was joined by Captain Jim Bingham who had brought a trenching spade. We soon had a proper foxhole dug. All hell broke loose that night, our guys were firing at anything that moved. Dr. Herbert (Herb) Pohle went to Jackman Pyre's foxhole to check on the status of the Signal Corps machine guns. He came back later and told me that they had been overrun. Exhausted, he asked me to keep watch as he took a nap.

Doctors Shapiro, Middleton, and Murphy of the 44th, who were supervising some of the perimeter groups, tried to ascertain the whereabouts of the Signal Corps machine guns. They talked to some of the Signal Corps platoon members who had retreated. The distraught men told them about the grenade that landed in their foxhole. They said that their lieutenant was killed when he put his foot on it. It blew up and he bled to death within minutes. But by risking his own life, he likely saved the others from the blast.

Chet Gjertson recalled the ass-chewing that the machine gun crew received back at the 44th's officer's tent, "Lieutenant Hurster from the 187th went ballistic when he heard about the abandoned Signal Corps machine guns. He told the men to smoke a cigarette and get their asses back out there!" But as the Japanese further infiltrated the area, they instead were told to stay and help defend the camp.

Captain Ray LaFauci recalls the battle,

Len Carhart and I were assigned to the rear of the perimeter on the evening of December 10th. This was considered one of the more vulnerable spots because of the expected threat from the paratroopers who had taken the airstrip behind us. They were presumed to be planning to rendezvous with their infantry comrades coming down from the hills in front of us. The likely route would be through the 44th General Hospital area. Instead of moving quickly, the Japanese delayed to celebrate with sake before making their approach. I recall the Japanese shouting out in broken English, "Medic, help me!", "Corporal of the guard, Corporal of the guard." No one from our foxholes cared to respond. Steady firing from our line, shooting at whatever appeared to move, must have given the appearance of a formidable force.

Walter Teague recounted memories of the events,

I was very busy on the night of December 10th escorting small groups of men from the center of the hospital area to positions on the west perimeter. As the Japanese started their flanking movement from left to right, we repositioned small reserve groups to increase fire power. A fresh batch of rifles opposed the enemy concentration and repulsed every charge. Our guys were extremely "trigger-happy." That was a blessing because the enemy must have thought that they ran into a full infantry regiment.

A foxhole was the safest place to be, as Japanese bullets were hitting the 44th's tents throughout the night. Dr. Peter Midlefart, a surgeon from Eau Claire, WI, manned a foxhole that was covered by a downed coconut tree log. He noticed Willard Arnquist, a tall and lanky MAC officer who was lying prone in a muddy foxhole, too shallow to conceal him. He warned him to dig a deeper hole!

Dr. James B. Bingham Jr. was manning a foxhole at the perimeter. He looked over the edge and out into the wallow to see if any Japanese were moving near. He was told by another officer to get down, that he was drawing sniper fire. A moment later, he was not wearing his steel helmet. A Japanese bullet had penetrated the helmet and knocked it off his head and out of the foxhole. Dr. Bingham was not able to move his limbs. He then passed out, unconscious from the shock. The medics positioned behind him came quickly under fire to his aid. He was taken to a blacked-out surgery tent where he was operated on, less than 50 yards from where he was hit.

Captain Bertrand W. Meyer (MD), a young surgeon from the

University of Wisconsin, was close to Dr. Bingham when he was hit. He observed that, "The bullet creased his scalp and temporarily paralyzed him." After the war Dr. Meyer was the staff surgeon at the Los Angeles General Hospital. Over 37 years he operated on many Hollywood notables, including John Wayne and Humphrey Bogart. Dr. John Sims, who worked on Dr. Bingham that night, also noted that, "The bullet had taken out some bone from his skull."

I had once asked Dad about his helmet. "Would it stop a bullet?" I asked. "Hell no," he replied. "It could mostly protect you from debris from a grenade or bomb, or keep a coconut from knocking you out." He was near Dr. Bingham when he was hit. "The bullet went right through his helmet. He was fortunate to be alive."

Dr. Edward A. Birge was supervising a group of enlisted men in a foxhole. He suddenly became aware of being shot, "I was wounded while lying on the ground near the big foxhole by the officer's tent. I had my arm in the air when I was hit." He described his wound in precise medical terms,

> The bullet must have been a ricochet since it only penetrated a couple of inches of muscle and stopped just at the edge of the brachial artery (the major blood vessel of the upper arm) and brachial nerve (the nerve that crosses the medial side of the brachial artery and lies in front of the elbow joint). After being hit, I gave my rifle to Major Daniels (also a doctor) and went to the first aid station under my own control.

Dr. Birge also described a close call for some of their University of Wisconsin alumni that night,

> A mobile surgical team, led by Dr. Bachuber, had arrived at our camp earlier in the day, and planned to set up in the area in front of us. Since he was also a graduate of the University of Wisconsin, he had called on Colonel Weston. Weston insisted that he stay within

the hospital perimeter. I hate to think what might have happened if they had been out in the rice paddy that night.

Captain Ray LaFauci tells the story of another type of casualty,

Warrant Officer Len Carhart and I were in our adjacent foxholes when a number of carabao (water buffalos) broke loose from their village pen and stampeded toward the hospital perimeter. Len heard this big noisy object speeding toward him in the dark. He opened fire with a full clip from his M1. The frantic carabao went down just in front of his foxhole, splashing him with mud. For a moment he thought the large animal was going to roll over in his foxhole. But, in a final act of desperation, it got up and headed for the wallow. It took some additional rounds from other trigger-happy defenders before falling dead. Unfortunately, the poor animal did not know how to say "piggly-wiggly."

The next day Colonel Weston was visited by a distraught Filipino gentleman who loudly mourned the loss of his animal. The carabao likely represented his main investment and was a vital contributor to his family's existence. A few days later, Weston directed Len Carhart to go to the Village Fathers and make a formal apology for the loss of the animal. LaFauci commented, "The owner was still distraught at the loss of the animal, but the village received a good supply of fresh meat."

December 11, 1944 - Aftermath

The 44th General Hospital's long night of peril came to an end at dawn on December 11th. Through the night the hospital had been under direct attack by Japanese combat infantry. The brave officers and enlisted men of the 44th stood together and repulsed every attempt to overrun the hospital. The men on the front line included doctors, dentists, psychologists, administrative staff, corpsmen, ward attendants, cooks, supply men, and drivers from the motor pool. Untrained in combat, they held their ground, returning fire, risking their lives with unwavering courage.

Why did the doctors of the 44th take up arms, risk their careers, and abandon the oath to "do no harm"? Was it purely to survive? Was it the hope of returning home to their families? Was it to protect their brothers, lying next to them in the foxhole? Had they been angered by the continual stream of young, wounded GIs brought into their care? They had seen too many young men buried in the shallow, water-filled graves on Leyte. Had the 44th been inspired by the courage and conviction of the Filipino people who endured the brutal Japanese occupation? They had witnessed the effects of torture, abuse, and rape on the Filipinos. They saw homes and lives destroyed. They had cared for the sexually-abused women, some in their early teens. They had cared for the orphaned, pot-bellied children suffering from malnutrition. They saw the innocent faces of the children, too young to know of any other life. Maybe like Private Ova Arthur Kelley of Norwood, MO, they "just got mad." We can speculate on the reasons they fought, but if they had not been adequately armed, the Japanese might have killed many of them. The Japanese soldier, it seems, had nothing to lose, and fought to die. The men of the 44th fought to live, having a lot to live for. They stood together as brothers protecting each other to the end if necessary. It seems that the words of Jesus describe it best, "Greater love hath no man than this, that a man lay down his life for his friends" (John 15:13).

Dad was the officer responsible for field reports. On the morning of December 11th, he and his staff surveyed the battle area. Their official count was 33 Japanese dead. No one knows how many additional dead and wounded were moved back by the enemy before daylight. Souvenir hunters had a field day collecting Japanese weapons and personal belongings. It was usual procedure to search enemy officers for papers that could lead to intelligence. Indiscriminate pilfering of the dead was discouraged, but also common.

Dad returned home with a Japanese officer's sword and a bayonet. Higher-ranking officers often had jewels embedded in the handles of their swords. The sword that Dad brought back was unadorned, indicating that it likely belonged to a Japanese NCO (non-commissioned officer). Dad noted that the Japanese paratroopers were easy to distinguish from the infantrymen. The infantry's time in combat and lack of resources had taken an obvious toll on them. Their uniforms were worn and the men were very thin. The paratroopers were in much better shape. They were also better armed, some with sub-machine guns, and their uniforms were in much better condition.

The 44th's Supply Sargent Richard Janes and Corporal Manuel Mathews were proudly photographed with a captured Japanese flag retrieved from the battlefield. Janes' daughter recalled her father's story of the battle and its aftermath, "With communications cut off, and the unit surrounded, they sat there like sitting ducks fighting off Japanese soldiers, all the while waiting for the U.S. military to come to their aid. But the U.S. military, believing that the men were dead, gave up the airstrip and notified the Red Cross that they were to notify the families back home. When the army decided to 're-capture' the airstrip, they were surprised to find that it had been under U.S. military control the entire time."

According to Janes' account, some weeks after the battle, an unfortunate Red Cross worker walked up and knocked on the door at 5th & Thompson Streets in Carson City, Nevada, to notify one soldier's wife that she was now a widow. Luckily, she had already received word from her husband that he had been to 'hell and back', but that he was alive and well. Janes' daughter commented, "When the Red Cross worker gave her the 'bad' news, I think she decided to send him to 'hell and back', too."

According to his daughter, Janes used to get a big kick out of the television show, *M*A*S*H*. He said it was very true to life. The 44th even had a "Corporal Klinger," only this guy called himself "Foot Locker." Whenever they would ask his name, he would respond, "Foot Locker." Janes' daughter recounted, "Dad said he could never figure out if the guy was genuinely crazy or just trying for a Section 8. But they finally ended up sending him stateside." Janes daughter added, "But Dad laughed the most at the episode where Radar O'Reilly is on the radio trying to get the friendly-fire shelling of the hospital stopped. He said that, at the beginning of 'Buffalo Wallow', when they still had communications, headquarters kept telling the hospital to quit firing. Because they must be shooting at their own men! Janes said they kept telling headquarters, 'These guys are a little too short and a little too yellow to be our men!'"

The planned counterattack by the Japanese paratroopers in combination with the 16th and 26th infantry divisions had failed. Aside from destroying small planes, supply and gasoline dumps, and a large number of Americans killed, nothing strategic was achieved. The paratroopers had surprised the ill-prepared American service units, but were routed when reinforcements arrived. The Japanese airborne units had only carried out five paratrooper assaults in the entire Pacific War, two of these targeted at Burauen. General Makino's famed 16th Infantry Division had been mostly wiped out in its final burst of offense on December 9th. The estimated 300 men of the Japanese 26th Infantry Division that had reached the Burauen airfields on December 10th made little impact. Few, if any, of the 26th Division survived the War.

On December 7th, Major General Andrew Bruce's 77th Infantry Division landed on the beaches near Ormoc on the west coast of Leyte. General Suzuki had to shift his focus to the U.S. landing at Ormoc which threatened his own rear positions. As Ormoc was essential to the survival of the 35th Army, he called off the counterattack and ordered all remaining troops to return there to defend his remaining supply lines. It was the last territory that the Japanese controlled on Leyte. With the loss of naval and air power, the Japanese lost their ground on Leyte. The Japanese logistical port at Ormoc was being surrounded by the 6th Army who hemmed them in on both north and south sides of the Ormoc Valley. They now faced a slow, but eventual defeat in the Battle of Leyte. Suzuki's 16th Infantry troops struggled back over the mountains to defend against the invasion. The rugged journey again took its toll. Those who made it back, returned as individuals. All unit integrity was lost in crossing the mountains to get to the Ormoc valley.

But there might have been a different outcome, given a different turn of events. What if the Japanese counterattack has been coordinated as planned? If the paratroopers had been supported by infantry units and had attacked in unison, they might have been able to hold the Burauen airfields until reinforced. Another two groups of paratrooper units were scheduled to arrive, but were forced to delay due to bad weather. What if the 44th had not obtained arms? Would the 26th Infantry have spared them and their patients? The 44th's officers were unanimous in their praise of Colonel Waddell for his persistence in requesting arms. Would other medical commanding officers have been as persistent? If not, what would have been the results? It's chilling to contemplate a different outcome. Fortunately, the 44th just suffered two non-fatal casualties in the Japanese attack. The worst thing that some suffered was a case of hookworm. As the 44th's foxholes were hastily dug, many contained contaminated soil from former village latrines. As the men came in contact with the soil, they contracted the parasite. In official records it's reported that Colonel Waddell was relieved of command late in 1944. The reason given was "due to illness."

Some members of the 44th criticized the Signal Corps platoon for abandoning their machine gun post when the Japanese attacked. Some of the 44th's officers looked at their actions as insubordinate. But in looking at the circumstance they were put in, I can't blame them for their actions. The lack of a larger infantry presence put them at great risk. In the darkness, they were ambushed and overrun by a superior Japanese force. They were faced with an attack they were not able to defend against. By joining with the 44th at the hospital's perimeter, collectively, they were stronger. Evidence is not clear, but members of the Signal Corps defended their actions by stating that their commanding officer gave them the order to retreat.

The enlisted men must have recognized the predicament they shared with the Signal Corps. As Christmas of 1944 approached, they evidently had some fun in retrospect. The 44th's medics and the Signal Corps staff traded humorous poems to joke about who was really responsible for the victory at Buffalo Wallow. In keeping with the spirit of the season, these parodied the classic "Twas the Night Before Christmas." The poems were shared by the Janes family.

The 44th's EMs (Medics) authored the first poem which was sent to the Signal Corps:

'Twas the week before Christmas
And all through the land,
The boys went to bed
With grenades in their hands.

The night was quiet
As our boys went to bed,
And we were thankful
To lay down our heads.

When out on the field
There arose such a clattering,
We sprang to our foxholes
As the bullets were spattering.

Our stomachs felt hollow,
Our throats were a lump,
And we kept our eyes
On the far distant stumps.

Our guns belched their fire
Across the mud hollow,
And history was made
At Buffalo Wallow.

Some Yanks on our left
Didn't care to fight,
As the Japs stared shooting,
They started their flight.

A phone call came next
From the man in command,
Why weren't we fighting,
He couldn't understand.

No Japs were around
At least so he said
But he came the next morning
To count the Jap dead.

The boys on our left
Had returned by this time
And you should have heard them
Giving out their line.

How they were the heroes
And that they did some brave deeds,
While we, of course,
Said our prayers on our knees.

Then a paper came out
With an editor's note,
About the boys on our left
And the guns that they tote.

A Presidential Citation
They must receive,
So, they can prove to their folk
They did a brave deed.

To us went their pity
Because of our plight,
You see we are medics,
And we surely can't fight.

So, let us pack our bags,
This very day
And send us back
To the U.S.A.

There we can tell
Of you brave boys over here,
While we love up your women
And drink up your beer.

We thank you, our heroes!

—Anonymous

The Signal Corps replied with a poem in response, called "The Heroes of Buffalo Wallow":

'Tis the day after Christmas
But we still remember,
That dark, dreary night
The tenth of December.

The boys were all tucked
In their G.I. sacks,
After a hard day work
They were flat on their backs.

When suddenly there came
From out of the dark,
The screaming of bullets
Trailed by tracer sparks.

The boys hit the dirt
In a hellova hurry,
The Nips were attacking
But we didn't worry.

Our plans had been made
Several days in advance,
The guards manned the perimeter
It was our only chance.

The Nips had machine guns
Perhaps two or three,
And I was scared stiff
(just between you and me.)

The Nips were a jabbering
To beat bloody hell,
We lobbed a grenade
Then heard one yell.

133

Two hours we battled
We knew we must hold,
Cause close on our right flank
Were men much less bold.

These men on our right
Were Medics you see,
Whose job is not fighting
But carrying pee.

They shine up the bedpans
And make up the beds,
So our heroic wounded
Can lay down their heads.

The battle kept raging
A runner came down,
The Nips had infiltrated
We have to give ground.

"To Hell with the order"
The G.I.s all cried,
Let's hold this damn line
Till the Nips have all died.

But orders were orders
The C.O. sent 'em down,
We had to withdraw
But not clear to town.

We formed a new line
The Nips there to hold,
To protect those on our right
Who we knew were less bold.

The Nips we had stopped
They advance no more,
The Signal Corps boys
Had exacted a score.

We advanced into camp
Next morning at light,
To drive the Nips out
With all our armed might.

And what did we find
My Medico lads?
A lot of good Nips
Cause they were all dead.

Nineteen there we counted
All huddled in heaps
But we didn't shoot 'em
Hell, we killed 'em with Jeeps.

Well boys, that's it
Now you know the whole story
Of how Signal Corps boys
Earned all of the glory.

One word of advice
Now take it from me,
You boys stick to bedpans
And carry the pee.

The 44th's EM (Medics) sent a final response to the Signal Corps:

'Tis four days after Christmas
And we, too, remember
That hellova night
The 10th of December.

The boys on our left
Were all tucked in their sacks
While we stood ready
To repulse the attacks.

The little Jap bastards
Attacked in the night;
We held our ground
Did they? That's the bite.

The Nips had machine guns
To help settle the score
And where did they get them?
From our own Signal Corps.

The boys on our left
Were so completely composed,
That they fled through our area
Without any clothes.

'Twas the CO's orders
For you to retreat,
But we can't account
For the many bare feet.

It must have pleased
Those Nips in your lines
To see all of you
With bare behinds.

You formed a new line,
Or so we were told,
But it was behind
Those Medics less bold.

We shined up our bed pans,
That we admit,
And we could have filled them
With Signal Corps shit.

And when you returned
To your camp at mid-day,
Everyone knew
'Twas your grandstand play.

You'd picked up some blankets
On your flight
And looked like Indians –
Oh, what a sight!

THE BATTLE OF BUFFALO WALLOW

You crawled along
Flat on the ground
And everyone knew
No Japs were around.

You'd best be quiet
And take it from me,
In spite of your claims
You'd make poor infantry

Keep stringing your wires boys,
And when the Nips roam,
You can call back from town
On your own telephone.

We are much less bold,
Or so they say,
But we'll take them on
Any old day.

—Anonymous

The gamble made by General MacArthur on Leyte trumped the hand of Japanese Generals Yamashita and Suzuki. MacArthur focused on the west coast of Leyte, leaving few infantry and an ill-prepared group of service personnel to face the Japanese counter-attack at Burauen. In my opinion, the 44th and the other service units were left to fend for themselves, expendable in the pursuit of MacArthur's timeline on Leyte. MacArthur continually pushed General Krueger, who led the 6th Infantry, to speed up the conquest of Leyte. Also, disturbing to me is the fact that the 11th Airborne's General Swing, in charge of the troops at Burauen, had received intelligence predicting the Japanese counterattack and infantry operations the day before the December 6th paratrooper drop. Higher-level officers dismissed the reports. When the events started to unfold, Swing knew exactly what he was up against. My opinion is that he too was "hung out to dry" by his superiors, who did not want to jeopardize the "strategic" actions along the west coast. I think that General Swing did the best he could with what he had, evidenced by the fact that he enlisted the assistance of any personnel he could find, and organized them into an opposing force. Thankfully, the ad hoc group of American defenders prevailed at Burauen. The Americans defeated the Japanese on the west coast of Leyte and the conquest of the Philippines shifted north to Luzon. MacArthur met his objective on Leyte and shifted the tide of the War. The Japanese, now severely crippled by the loss of Leyte, would have to dig-in to defend their remaining positions and their homeland.

The Battle of the Airfields

To the men of the 44th, their encounter with the Japanese infantry on December 10, 1944, was dubbed the "The Battle of Buffalo Wallow." Maybe someone of the 44th had studied the history of the American West, as this was also the name of a battle that took place between U.S. Cavalry and Native Americans in 1874, as part of the Red River War, fought in Texas and Oklahoma (refer to the description of this event at the end of this book). Historically, the events in late November and early December of 1944 at Burauen, Leyte, have been referred to as "The Battle of the Airfields." Listed below are some accounts from historians, official military journals, and war correspondents.

From *Hirohito's War*, Francis Pike, (London: Bloomsbury, 2015), pp. 942–944, referring to the first Japanese paratrooper raid that failed:

> *The 2nd Parachute Brigade launched a suicide mission on the night of November 27, 1944. They planned to seize the airfields in a night-time raid and hold them until an assault by General Suzuki's 35th Army's 16th and 26th divisions.*

THE BATTLE OF BUFFALO WALLOW

From *Leyte 1944*, p. 232, referring to the fighting on December 10th, and the 44th:

The attack pushed the defending Air Force personnel back until they reached the hospital area, where they held. A counterattack soon drove the Japanese off, leaving over thirty dead behind them. The Japanese Army's major counterattack on Leyte was finally over.

From *Leyte 1944: Return to the Philippines* – describing the Japanese counterattack:

Although the Japanese counterattack on the Burauen airfields caught the Americans by surprise, it failed to retain total control of Buri or San Pablo. Americans at both airfields reorganized their forces and defended their positions. A mixture of engineers, artillery gunners, 11th Airborne Division glider infantry, 38th Infantry Division units, and a tank battalion retook the San Pablo airfield on December 11th. No Japanese paratroopers survived. At Buri, Japanese forces tried to push forward to Bayug, but there were no supporting forces to assist them as planned. The 16th Division was decimated and the 26th Division did not reach Burauen in adequate strength. Japanese forces, without supply, ran out of ammunition and had to rely on what they could capture from the Americans. Suzuki ordered the remaining 16th Division troops to combine with the 26th and retreat. The combined "Wa" and "Te" operations had failed. With the U.S. 77th Infantry Division landing at Ormoc to the west, the last Japanese-held territory on Leyte was in jeopardy.

Pat Murphy of the *Wisconsin State Journal*, Washington Bureau posted the following story, titled "44th State Hospital Unit Praised for Medical Service – And Fighting," in early 1945:

WASHINGTON - Somewhere in the maze of islands constituting the Philippines is the only all-Wisconsin unit in action in this war on any front. It is the 44th General Hospital unit, which was recently visited by Major General Norman T. Kirk, Surgeon General of the Army. Consisting almost entirely of medical personnel from Wisconsin General Hospital, Madison, the 44th trained and organized at Ft. Sill, Okla. Part of the first group left for Queensland, Australia in the fall of 1943.

Prove Good Soldiers Too – "These men," Gen. Kirk told the State Journal, "were doing a grand job and their hospital was one of the best I visited. They were not only rendering splendid medical service to our wounded men but proved themselves good soldiers in defending their hospital against the Japs."

By late fall of 1944, the 44th was in the Philippines, moving inland to a supposedly quiet area. They found themselves at the end of the line, literally in front of portable hospitals and collecting companies. In front of them were rice paddies (wallows for the water buffalo) and rugged hills and low mountains.

And in the hills? They soon learned that "thar were Nips in them thar hills," according to the graphic description of one of the members of the unit. The following account given by that unidentified member, ties in nicely with Gen. Kirk's reference to the ability of the 44th to defend the hospital against the Japs.

Early in December the Jap paratroopers dropped right behind the hospital area. Doctors and corpsmen went out to get the casualties in neighboring units in the pitch dark – brought them in to be operated in the blacked-out surgical tent where two tables went for 36 hours continuously. Thus, the unit acted as aid-men, litter-bearers, collecting company, and give definite surgical treatment. That's one for the Surgeon General's book.

Medics Rout Japs – But the big event came about four days later when the Japs from the mountains assembled to join their para-troopers (a bit late for the airborne Nips were sopped-up quickly). They hit the perimeter manned only by doctors and corpsmen. A signal outfit on the left flank retired. A few infantrymen, less than a squad, wandered in and were put to work.

One of the doctors who had helped on the NP service was in command of the defense. From collective descriptions it was a night-mare. Torrents of rain – pitch darkness. The Nips captured a few machine guns from the outfit that retired and fired the tracers into the hospital area all night. Snipers abounded. The usual Nip night tactics were used – shouting, yelling, calling for medics, infiltration, throwing grenades after drawing fire.

Two of the medical officers were wounded in defending the hos-pital, rushed a scant 100 yards to the surgical tent, were operated on and revived on the spot.

At any rate, the Jap attack was fought off – and nary a man was lost. Dawn patrols revealed many Nips dead, and the souvenir collectors had a field day.

Whereabouts Secret – After that, the 44th had infantry protec-tion, but the Nips' patrol attempted to infiltrate nightly, employing their usual fiendish tactics. Many of the medics slept fully-clothed.

Where the 44th is just now cannot be revealed. The job of reclaim-ing the Philippines is getting on and no doubt the 44th is going calmly about the business of saving human lives.

Doctors and nurses are modest and unassuming, and while not a great deal has been said concerning them, perhaps it is because the glory of their work cannot be limited by words. Civilians do not like to be ill, but the fear of illness is definitely minimized because in childhood, faith that the "doctor will make you well" was instilled.

Probably no soldier believes he will be ill or wounded. But if he is, he knows that he will find himself in the capable hands of doc-tors and their assistants who will make him well again… men and women exactly like those in the 44th General Hospital Unit, a few yards back of the lines, just waiting to do that job.

An article was published in the *Philippine Post*, titled "Battle of Buffalo Wallow" on December 11, 1944, one day after the battle. This dispatch is referenced in the poems traded between the 44th's EMs and the Signal Corps platoon referenced below. Note the discrepancies between the original dispatch, the editor's note, and the last paragraph (added later by someone unknown, and not part of the published version), refuting the story and aligning it with the 44th's veterans' accounts:

FRONT LINE DISPATCH: 11 December 1944 – While the American 77th Division troops were battling desperate Jap resistance at the gates of Ormoc, history was being made in Central Leyte; perhaps the most amazing battle of the war was being fought, and won, without firing a shot. This startling action, known as the "Battle of Buffalo Wallow," took place in the black of night, along a treacherous swamp, when a Jap patrol, armed with automatic weapons, infiltrated to the edge of the swamp. It was a tight fix for a service group, dug in on the other side of the mire, armed only with semi-automatic weapons. In the teeth of withering enemy automatic fire, the Yanks seemed to have nerves of steel; more amazing, this was their first experience under fire. Showing vise-like control, our boys didn't fire a shot, didn't give away their positions; while the Japs, having shot-up all their ammunition, had lost the "Battle of Buffalo Wallow."

EDITOR'S NOTE: We appreciate the fact that a certain signal company wasn't able to hold its fire, mowed down a score of Japs, and as a result probably saved many lives in a certain hospital area. We know there were other service outfits that were on the ball during the "Battle of Buffalo Wallow." Our hats are off to all of you. As for the Infantry, good work and just remember that soon everybody *goes home on rotation (it says here in small print) and the more of those ***** we knock off, the quicker we go home to enjoy the better things in life; need I go into detail?*

WHAT REALLY HAPPENED! - On the night of 10 December 1944 an unknown number of Japs infiltrated into the area occupied by the 44th General Hospital and the 440th Signal Group. With no combat units in the immediate vicinity, it was necessary for the untrained Yanks of the 44th, with the assistance of the Signal Group, to defend their perimeter. This being their baptism of fire, the boys of the "Fighting 44th" gave an excellent account of their ability to cope with any situation. The Signal Group put up a mighty swell fight, but after several hours of fighting were forced to retreat, leaving the battle to be finished and won by the "Fighting Medics." Much fierce fighting was done, with an unknown amount of ammunition being used (and that was plenty) by both Japs and the Yanks. Fighting continued on until near day-break. The net results of the battle were thirty Japs by official count, who went to meet their Honorable Ancestors. And that, friends, is the real story of the "Battle of Buffalo Wallow."

The official U.S. Army records described the events of early December at Burauen:

Some 350 Japanese paratroopers dropped at dusk on 6 December, most of them near the San Pablo airstrip. Although the Japanese attacks were poorly coordinated, the enemy was able to seize some abandoned weapons and use them against the Americans over the next four days. Hastily mustered groups of support and service troops held off the Japanese until the 11th Airborne division, reinforced by the 1st battalion, 382nd infantry, and the 1st and 2nd battalions, 149th infantry, 38th infantry division, concentrated enough strength to contain and defeat the enemy paratroopers by nightfall of 11 December. Although the Japanese destroyed a few American supply dumps and aircraft on the ground and delayed construction projects, their attacks on the airfields failed to have any effect on the overall Leyte campaign.

From official Army Medical Corps records, Colonel Wm. Barclay Parsons, MC, the Surgical Consultant in USASOS (the U.S. Army Services of Supply), SWPA (the Southwest Pacific Area), documented visits on Leyte for the U.S. Army Medical Department, Office of Medical History. His log notes provide the most detailed account of the 44th's experience in the official records. On December 19, 1944 he logged this report:

Saw Colonel Weston, 44th General Hospital. They were set up near an airstrip. About 600 Japanese paratroopers landed on the strip, and a few nights later reinforcements came in to join them. Our men decided to hold the perimeter, as they had about 200 patients in tents and the road in was impassable. The Japanese came into a Signal Company first, and the guards were ordered out to the perimeter. By this time, the Japanese had machine guns set up on three sides of the hospital. There was an all-night fight, and in the morning, they found 23 dead Japanese. Two officers of the hospital were wounded. None of the hospital personnel had had training in firearms other than squirrel shooting. The commanding officer suggested before leaving the States, that they should have such training, and he was told that all the training they would need was in getting into formations, so as to be able to march on and off the trains.

The consultant's report was continued on December 20, 1944:

Today, I tried to drive to the 44th General Hospital, but the bridges were still out and the roads were impassable. D+60, and no other general hospital has been set up as yet to receive surgical patients. Visited the 116th Station Hospital which never took any patients and is now moving.

The Department of the Army, in 1948, published a thirty-volume set of books entitled *The U.S. Army in World War II*. On page 309 of the volume describing the Leyte campaign, there is a photo of burned out P38 planes on the Buri airstrip. It just states that the airstrip had been mostly abandoned because the roads were impassable in the area. There is no mention of the Japanese paratrooper attack, the Army Air Corps personnel who were killed, or the fact that there was an active General Hospital just a few hundred yards west of the airstrip. And that the hospital was subsequently attacked by Japanese infantry on December 10, 1944. The 44th was possibly inferred in a reference to the "support and service troops" that engaged the Japanese during their Leyte counterattack. Daily field reports have also disappeared. It seems that facts were conveniently erased from the official Army records. It's as if the experiences recounted by the 44th's veterans didn't happen.

Even in the many recent WWII history books published, the 44th General Hospital is not specifically mentioned. The abandoned machine guns are mentioned in some historical accounts. A hospital is mentioned, but not specifically the 44th General Hospital. Only Colonel Parsons, MC, (referenced above) and the Surgeon General of the Army, Norman Kirk, mentions the fact that the doctors and corpsmen of the 44th, untrained in combat, armed and fought off a Japanese attack. Kirk's article, published in *Collier's* magazine, is presented below.

That They May Live!

A direct attack on a U.S. Army hospital was in gross disregard of the Geneva Convention. During the Pacific War, the Japanese were known to shoot medics, shell hospital camps close to the lines, and even attack U.S. hospital ships at sea by air. But a direct infantry attack on a general hospital was unheard of in all of World War II, even given the known brutality of the Japanese and Germans.

Unapologetic, the Surgeon General of the Army, Norman T. Kirk, praised the actions of the 44th General Hospital in a *Collier's* magazine article. In the article, the highest-ranking leader of the Army Medical Corps, skirts the issue of arming the hospital staff. The following is from the *Collier's* magazine article, "That They May Live!" by Major General Norman T. Kirk, The Surgeon General of the Army, dated July 21, 1945. I've added some comments in brackets:

> *One of the American field hospitals in the Philippines recently was moved into what was supposed to be a quiet sector* [the 44th]. *But when the medical officers and corpsmen arrived, they found themselves virtually in the laps of the Japs. The Nips waited until our boys settled, then one night they dropped a welcoming party of paratroopers right behind the hospital area* [the December 6th evening attack on the Burauen airfields].
>
> *All the Japs were killed, but the medical detachment suffered casualties, too, and ran a little low on manpower. So, doctors and hospital corpsmen went out and brought in the wounded and oper-ated on some of them in the blacked-out surgical tent where two tables were in constant use for 36 hours. Thus, for a bitter and perilous period, they functioned as first-aid men, litter bearers and a collecting company when they weren't busy with surgery. After it was over, they said, "Now we have seen everything."*
>
> *They learned otherwise four nights later* [December 10th] *when Jap infantrymen came out of the mountains* [the 26th Infantry] *to join the airborne soldiers who, of course, were dead. Only phy-sicians and corpsmen, doubling as doughboys, were available to man our perimeter in defense of the hospital. Heavy rain made the night so black that our men could barely see the weapons they held. The Japs used their night tactics of yelling, calling for medics, and infiltration, while their machine gunners and snipers peppered the*

hospital area. As infantrymen, our medics did all right. They beat off the attack, killing many Japs, with no fatalities to themselves. Two medical officers were operated upon and convalesced less than 100 yards from where they were wounded [Dr. Edward A. Birge and Dr. James B. Bingham, Jr. of the 44th GH].

We of the Army Medical Department are proud of the fact, as demonstrated by this story, that our men can give a good account of themselves as combat soldiers [even though the 44th was not explicitly trained in combat prior to leaving the States]. *But we are prouder of the record we have made in our own job which is not to fight, but to spare men as much as is humanly possible from the disfiguring, maiming, killing business of war. Mothers of our fighting men will find more comfort in the knowledge that our medical officers know how to heal, than in learning that they also know how to fight.*

We do not, of course, make a practice of maintaining field hospitals right in the front lines. We put our hospitals, however, as close to the fighting as possible [IMO, a bit too close for the 44th]. *Our field hospital platoons, supplemented by surgical teams, operate from two to four miles behind the forward lines. We have brought surgery up to the patient. In fact, medical care begins almost as soon as the soldier becomes a casualty.*

This glowing account by the Army Surgeon General obviously shows his pride in the performance of the 44th. At a critical time, the 44th functioned as a combat unit and a hospital, in spite of the fact that they had not received any weapons training and were officially denied the acquisition of arms. This had little precedence for a General Hospital in any war.

Walter Teague, the 44th's Mess Officer, completed a distinguished career with the Army Medical Corps, serving in the Japanese occupation, the Korean War, and Vietnam. He recalled his first assignment stateside, in 1946. "I was sent to Ft. Benning, GA. My Commanding Officer was Brigadier General Robert Hill, MC. He had served with General Kirk in the

Philippines before the War. General Kirk visited Ft. Benning during the summer of '46 and gave a glowing talk about the 44th in the Philippines. After his talk he noticed my shoulder patch and that I had served with the 44th. He again expressed great pride in our unit and showered me with questions. I had assumed that all that he said about us was published, but sadly not much survived."

As stated previously, Article 19 of the Geneva Convention states that medical units, i.e., military hospitals and mobile medical facilities, "may in no circumstances be attacked." There are also provisions defining what medical personnel can do in "lawful self-defense" vs. engaging in an "act harmful to the enemy." Determining the conditions that forfeit a medical unit's protected status is a topic still debated to this day. It appears that the reason for the removal of information about the 44th General Hospital's actions had been the desire to demonstrate adherence to the Geneva Convention. The Convention had not been respected by the Japanese. The question then becomes, who was the first to abuse the Geneva Convention, the Japanese for attacking, or the 44th for arming and defending?

But I think that there's a more fundamental question to ponder. Was it a tactical mistake by the Army that left the 44th on the fringe of an active combat zone without protection? Or, was it a strategic decision, driven by the desire to advance the timeline on Leyte? Perhaps available infantry were purposely not diverted, so not to risk jeopardizing MacArthur's bold move to surprise the Japanese at Ormoc. Were the American casualties at the Battle of the Airfields conveniently couched in the overall battle for Leyte? According to some of their veterans, records of the 44th General Hospital received a SECRET stamp at the end of the War. It seems that either a tactical blunder or strategic choice prompted a cover-up to protect higher-level commanders.

The Final Act

One Saturday morning I went shopping with Mom and Dad. Before the large grocery chains took over, we had a small grocery store just up the street from us, called Wayne's Market. Wayne, the owner, was a large man with black glasses and a military haircut. Wearing his white butcher's smock, he worked between the meat department and the front checkout area. Wayne always took time to interact with the people in the neighborhood who frequented his store. He and Dad would have some lively conversations and he always had flattering complements for Mom. As a boy, I wore a green Army jacket. Mom had sewed Dad's Sergeant's stripes on the sleeves, along with some other insignia. Wayne would greet me with a salute and say, "Hello Sarge! How's the war going?" He and Dad were amused as I'd clumsily return his salute. I suspected that Wayne was also a veteran.

As we walked through the produce section, I noticed the raw coconuts stacked in a pyramid. I always wondered what eating a real coconut would be like. They were curious things. It had three small, flat round areas on the top, looking like the areas where you'd drill holes on a bowling ball. As you held the rough shell and shook it, you could hear the coconut milk sloshing inside. I wondered how you opened one, as the shell was hard and the surface rounded. As Dad saw me looking at the coconuts, he asked if I wanted to try one. I eagerly selected one and placed it in the cart.

That evening I sat on the patio with Dad. I asked him if we could try the coconut. He said that he'd get what we needed to open it up. I went into the kitchen and grabbed the coconut. Dad returned with a knife and a wine bottle corkscrew. The knife he returned with was a Filipino bolo knife that he had brought back from the War. The knife was housed in a coconut wood sheath that was cut into two identical pieces. The two parts were tied together with thin, but strong strips of dark brown leather. The knife's handle was carved in a diamond shape. The blade was thin, sharp, and curved like a "lazy" letter S.

Dad said that as they moved across the Southwest Pacific, they would pay native boys to climb the trees and knock down the large coconuts. Part of this was for safety reasons, as Dad said that a coconut hitting you on the head from twenty feet up could knock you out if you weren't wearing a helmet. Also, the coconut milk was a great alternative to the chemically-treated water they drank. Dad took the corkscrew and opened up two of the holes on top of the coconut, then drained the coconut milk into a glass. Then he took the bolo knife and striking the coconut on its side, split it perfectly in two. The insides were nothing like the sweet, dried, shredded coconut I was familiar with. The thick, raw coconut slices took some getting used to. The watery milk, was OK, but also not particularly sweet. But, as Dad explained, if you were on an island and didn't have much else to choose from, the coconut could be a lifesaver.

Dad explained that the Filipinos used the bolos, or machetes, for many things. The knife could cut through jungle growth on a trail, be used in harvesting foods, and also used as a weapon. I believe that the knife was a gift from one of the Filipino guerrillas on Leyte that he had become acquainted with. I could tell that it was a special memento from the War.

Dad didn't talk about specific Filipino people he met on Leyte. But it was evident that the Filipino people had a profound effect on Dad and other members of the 44th. Out of the many photographs taken by Dad during the War, the majority of them were of the Filipino people. He took photos of them going about their daily lives. He took photos of groups of children, many with a look of despair and uncertainty on their faces. That was also true of fellow MAC officer Walter Teague. Many of the photos he took were of Filipino people in scenes of everyday life. Some were of children and family groups who smiled broadly at the camera. I assume that the two amateur photographers knew the villagers well.

The 44th obviously shared a common enemy with the Filipinos. Dad was always complementary of the fighting spirit of the Filipinos. He knew some that were part of the guerrilla outfits who fought the Japanese in the hills, jungles, and swamps of Leyte. He also spoke of Filipino scouts who alerted the 44th of Japanese movement and activities in the area.

The 44th also depended on Filipino laborers who did the hard work of digging ditches and foxholes, and other construction tasks. The medical team benefited from the people who helped them with transporting the wounded, and assisted with some of the manual work of running a hospital. The mess officers benefited from using the fresh eggs and tropical fruits that the villagers would harvest and sell. Village women provided laundry services for the troops. They would launder a uniform for a single peso. They would typically beat the clothing in the local river, then dry them in the sun or, during the rainy season, over wood fires. The 44th personnel appreciated having clean clothes, free of mud, but talked of smelling like a campfire. Money made cleaning or doing work around the camp was sometimes gambled on Sunday cockfights by the Filipino men.

From the photos he took, I imagine that Dad was particularly touched by the condition of the Filipino children, who had no control over the situation they were born into. Many children were orphaned during the Japanese occupation. Some may have witnessed their mothers, fathers, and siblings killed or brutalized by the Japanese. Veterans of the 44th described situations where resistant Filipinos were ordered to dig their own graves, then were shot or beheaded on the spot. The children experienced hunger, brutality, the sound of battle, and the stench of death. I expect that I'm unable to fully contemplate the extent of the suffering and emotional impact.

But in spite of the hardships, Filipino society provided a glimpse of home for the U.S. troops. Their families were close-knit. They were good at working the land. They worshipped together in familiar churches. In spite of the wartime conditions they carried on with life, enjoying parades, festivals, and sporting events.

The Filipinos benefited from the partnership with the U.S. troops and the 44th in particular. Besides the security offered by the Army's presence, they were grateful for the medical staff. Although the Army intended that Filipinos be routed to civilian hospitals, many needed immediate and/or advanced care. Many Filipino citizens were brought to the 44th General Hospital, some who had been bayonetted and left for dead by the Japanese.

Young women who had been repeatedly raped and beaten were brought in. As were sick children, who had received wounds from battles, and/or contracted disease from the harsh conditions. Doctors and nurses, when they had time off, would sometimes make "house call" trips into Burauen village to provide medicine and care for the civilians. The Filipinos, in their gratitude, would send them back with tropical fruits and local foods.

The Filipinos suffered great loss of life and tremendous physical destruction during the War. By the end of the War, an estimated 527,000 Filipinos, both military and civilians, had been killed from all causes; of these between 131,000 and 164,000 were killed in 72 war crime events. But the smiles seen on the faces of the people Dad photographed, of a people who had been through such horrific times, seemed to offer a ray of hope for a better world in the future.

The 44th General Hospital was transferred to the 8th Army on December 25, 1944. On December 27th, General MacArthur announced that the Japanese Army and General Yamashita had experienced their biggest defeat of the war. He described the planned next steps on Leyte as "mopping up operations." This proved to be a premature announcement, as it would take until the end of the War to drive the Japanese completely off Leyte. MacArthur assigned the "mopping up" task to the 8th Army under General Eichelberger. The 6th Army, led by General Krueger, would proceed north to invade the largest Philippine Island, Luzon, and the capitol city, Manila. This would be another arduous invasion and campaign. An even higher number of casualties would start to flow to the 44th and other hospitals.

The 44th's nurses arrived in January of 1945 from New Guinea. By that time, the area was considered safe, although Japanese were still active in the area. It must have been an interesting reunion, as the men of the 44th told their stories of their experience in defending the hospital. The doctors, corpsmen and patients, were glad to have the nursing staff back to handle the patient load.

On April 12, 1945, it was announced that President Franklin Delano Roosevelt had passed away. America and its troops around the world mourned his passing. FDR had been their president for close to twelve years, leading the U.S. through the dark times of the Depression and into World War II. Most people were skeptical that Vice President Harry S. Truman would be able to step into the role, particularly at this critical juncture of the War. Many were not even familiar with the man from Missouri. But Dad knew who he was, as he had seen him many times in Kansas City, both as a shop owner and political figure.

Better news soon followed as Germany announced the unconditional surrender of its armed forces on May 8, 1945, marking the end of World War II in Europe. Celebrations resulted around the world. Troops in the Pacific hoped that they would soon receive the additional resources they needed to finish off the Japanese and return home.

The 44th remained at Burauen through May of 1945. When General MacArthur's headquarters moved from Tacloban, Leyte, to Manila, Luzon, the 44th moved into their abandoned buildings. The complex consisted of several administrative buildings, a large Officer's Club, and a mess hall with views of the beach. The facility at Tacloban was a relative "tropical resort" compared to the months at Burauen. Wood floors, not mud. Running water, electricity, and indoor plumbing. Screens on windows, and a breeze off the water of the bay, kept the bugs in check. A broad, sandy beach was a short walk from the facilities. The beach would be a popular place for getting some much-needed R&R.

But there would be limited rest time for the 44th. Casualties from the Luzon campaign and an increase of battle fatigue and tropical disease cases would add to the patient count. Japanese General Yamashita had left Manila and retreated into the mountains of northern Luzon, where once again he dug into fortified positions. The U.S. would experience horrendous fighting there, trying to dislodge the dug-in troops. Yamashita, who earned the moniker "The Tiger of Malaya" in his homeland, was now referred to by U.S. troops as "The Gopher of Luzon."

The nurses of the 44th were an exceptional group. The stories shared by nurses of the 44th, including Colonel Ida Bechtold, Captain Eda Teague, and Captain E.F. Hastings, are worthy of their own books. They described the challenges of caring for patients, many with horrific wounds. As the casualties increased the patient count to over 1,500, they worked long 12-hour shifts to maintain 24x7 coverage. Those who worked during the night had it particularly rough, as sleeping during the day in the tropical heat was nearly impossible. The nurses also faced the same threats of Japanese snipers, shelling and strafing. They also succumbed, at some point, to the various tropical diseases they were also treating. Facing the dangers of war became routine for them.

A nurse's face would sometimes be the first thing that many wounded men would see when they were revived or came out of surgery. Many young men took comfort in having a female caregiver, reminiscent of their mothers and grandmothers. The nurses faced the challenge of comforting many who were not expected to live. They would take the time to write down their last thoughts and wishes in letters to loved ones. As wounded men slept under mosquito nets at night, some, upon hearing shelling nearby, would panic, wrapping themselves up in the nets. They would then be cut free by a nurse who would help calm them down. The nurses stated that knowing they had a critical job to do, kept them going in spite of what they experienced. They also knew that they could not get emotionally attached to those they treated, as this would take an even larger toll on them.

Some of the soldiers who died in the hospital would receive letters from home that would now never be answered. Dad and other officers would spend time writing condolence letters to family. In spite of the circumstances of a soldier's death, whether from battle wounds or from disease, they gave glowing praise for the loved ones' service and bravery. This provided some consolation for those who waited at home. Many families wanted to know the exact circumstance of the soldier's death; but many of those with them may have also been killed, or did not want to talk about it after the War.

As the heaviest fighting moved north to Luzon, the 44th was visited by many celebrities who performed at USO shows and stopped at the hospital to brighten the spirits of the wounded. Those visiting the 44th included Bob Hope, Irving Berlin, Jack Benny, and Danny Kaye. The future baseball

Hall-of-Famer and N.Y. Yankee shortstop Phil Rizzuto also visited. Rizzuto, who was serving in the Navy, joined in on a game at the 44th's baseball field. In the summer of 1944 comedian Bob Hope hopped from island to island in the South Pacific entertaining the troops and visiting hospitals. It was an emotional, as well as dangerous, journey of over 30,000 miles for Hope and the other performers who joined him. Hope would continue entertaining the troops through the decades, including making multiple trips to Vietnam in the 1960s and 1970s.

On the Homefront

I was recovering from the chicken pox during the spring of my 3rd grade year. After I'd spent four days at home, Mom was ready to send me back to school on Friday. But a beautiful, warm spring day changed her mind. It had been a typical cold and rainy early spring in Kansas. But on this day, there wasn't a cloud in the sky. The air in the backyard smelled like Mom's freshly washed linens, hung out to dry on the clothesline. Trees, flowers and birds seemed to come to life in the warm sunshine. She said, "I think that the sun will be good for you."

That sounded great to me. I wasn't ready to go back to school.

She added, "Let's walk down to buy some honey, then we'll come back and make some tea."

I looked forward to some time with just Mom and me. She was usually busy with the garden or housework in the morning, helped out at her parent's store at lunchtime, then back at home in the afternoon to start dinner. We walked down the road to the beekeepers' home. They were an elderly farm couple who had sold off some of their land to developers. Now they had a small orchard, a vegetable garden, and a neat row of white hive boxes. They had a box on a post next to their mailbox with jars of honey for sale. Through the glass on the front of the box you could see if there was honey available. An empty mason jar sat on one end of the box to collect cash and provide change. There was little threat of theft; in the 60s the honor system still prevailed. As we were paying for a jar, the beekeeper walked up and greeted us. He took us to the back porch where they had a beehive encased in clear plastic. I was fascinated to get this close to the bees and watch them at work. The man explained the workings of the hive. Worker bees gathered pollen, nurse bees converted the pollen into food to feed the queen bees larvae, while others defended the hive from attackers.

When we returned home Mom and I sat in the sunshine on the patio, enjoying toasted English muffins with butter and honey, along with a cup of black tea with milk and honey. Mom would occasionally talk about her experiences of the War. I asked her if she was ever afraid that Dad would not make it back. She said that they all kept very busy while Dad and her two brothers were overseas. But there was a certain time in the day when everyone would be anxious. She told me of the Western Union courier that would come down the street every day around noon. Mom's parents' home was attached to their store on 7th Street, a broad street that ran north and south through the heart of Kansas City, KS. To the north were the city government and postal services. The courier, in a black Ford sedan, had a Western Union sign on each door. He would drive up the broad street, sometimes slowing down to look for house numbers, other times stopping at a home. During World War II, many of the telegrams the courier delivered did not announce good news. The d readed w ords t hat n o o ne w anted t o hear was, "I regret to inform you…" Mom said that if he stopped in front of the house, her heart would stand still. Occasionally he'd stop at the store to buy a sandwich or grab a Coke, then shoot the breeze with my Grandfather and the old Polish men who congregated in the store. Mom was always relieved when the courier drove off.

In January of 1945 news spread of the loss of a good friend. He had been killed in the Battle of the Bulge. In the close-knit Polish community, they had grown up together, attending the same church and school, and meeting at the Saturday night Polka dances. They had celebrated at each other's weddings before the War. The women gathered baskets of food to console their friend, now widowed. In the front window of their home hung a small flag with a gold star. The reality of the War would hit many times in the community.

THE BATTLE OF BUFFALO WALLOW

In May of 1945, Mom had just received a letter from Dad telling her that things were fine and that he had moved to a better facility in the Philippines. He was happy that the War in Europe was near its end, and he hoped that more troops would soon be coming to the Pacific. He couldn't say where he'd be going next. No one expected the Pacific War to end soon. As the fighting on some of the Japanese home islands, like Iwo Jima and Okinawa had proven, the enemy was determined to fight to the end. "Golden Gate in '48" became a popular slogan. Dad asked Mom to send the latest news and photos of Eddie, now going on two years old. On a nice May morning, Mom sat on the front porch as the baby played with a set of wooden blocks. She was engrossed in responding to Dad's letter and did not hear the Western Union courier approach. He stopped in front of the home and got out of his vehicle. But instead of going into the store, he came up the steps where she was seated. She said that her heart sank and she started to tremble. He asked if her mother and father were available. As they stood sombrely on the front porch, they were informed that her brother Frank was missing in action. His destroyer, the U.S.S. *Little* (DD 803), had been sunk off the coast of Okinawa. They could not yet confirm his whereabouts. Mom's Mother clutched the crucifix on her necklace and quietly gasped, "Boże kochany!" ("Dear God" in Polish).

The U.S.S. *Little* was a Fletcher-class destroyer officially launched on May 22, 1944. The U.S. Navy's destroyers were the workhorses of the Pacific War. It was effective in many ways, bombarding the islands prior to landing, escorting transports across the Pacific and LSTs to their landing areas, scanning for submarines, firing at attacking planes, and performing "picket duty" offshore. The destroyers were equipped with sonar and the latest radar equipment, recent technology innovations. As the "eyes and ears" of the Fleet, they could warn other ships of incoming Japanese fighters and bombers while at their "picket" positions. Mom's brother Frank was a radio operator on the *Little*. Being positioned in a tower on the ship provided him with a birds-eye view of the ocean and activity below.

The *Little* departed from Seattle on November 11, 1944, escorting a convoy to Pearl Harbor. Commander Madison Hall, Jr. was her captain,

158

and Lt. Marlin (Spike) Clausner, the executive officer. There were 16 officers and over 300 enlisted men aboard. After 12 days at sea, the *Little* approached the surreal scene at Pearl Harbor, where the War had begun three years earlier. The ship blew a shrill eerie whistle as the men on deck stood motionless and gazed at the wrecked vessels visible in the water below. Many got choked up when they thought of the bodies of those who lay entombed in the water below. Patches of oil could still be seen along the surface of the otherwise clear and peaceful waters. It was a sobering reminder of the risks that lie ahead for them.

At Pearl Harbor, the *Little* participated in training drills for the invasion of Iwo Jima. The crew was allowed some liberty time in Honolulu before receiving their orders. It was strange for the men to spend Christmas away from home in this warm tropical paradise. The orders arrived, and on January of 1945 the *Little* was sent from Hawaii into battle. Her first stop was in the Marshall Islands on Eniwetok, captured in a short but hellish battle by U.S. Marines a year prior. After a long trip across the Pacific the captain gave the men some R&R time on this small coral atoll. A BBQ and beer party was held on the tropical beach. Men swam in the surf and in a clear blue lagoon. The pristine island would later be covered with radiation after the War. During the Cold War period, the remote island of Eniwetok would be the site of over 40 U.S. nuclear bomb tests.

From Eniwetok the *Little* departed for Saipan, in the northern Marianas Islands. The men were once again allowed to rest and go onto the island, but were warned of the existence of Japanese snipers. Although Saipan had been captured by the U.S. in July of 1944, many Japanese soldiers eluded capture and held out in the mountains, some until many months after the end of the War. Saipan had been a brutal three-week battle that resulted in more than 3,000 U.S. deaths and over 13,000 wounded. Over 29,000 Japanese soldiers died there as well as a large number of Japanese civilians who lived on the island. In horror, U.S. servicemen watched helplessly as over 1,000 civilians committed suicide by jumping off the highest cliffs and into the ocean. The civilians had been falsely warned by their military leaders that they would be brutally tortured by U.S. troops if captured. Saipan had been a key victory for the U.S. It resulted in the resignation of Japan's Prime Minister, Hideki Tōjō, and put the Japanese main islands within the range of U.S. B-29 bombers. The crew of the *Little* witnessed large formations of B-29s leaving in a V-of-Vs pattern in the morning, then return in the late afternoon. Sometimes the men would count the number that left to compare with the numbers that came back. Many never returned. Some returned severely shot up and damaged. B-29s that approached the landing strip with their lights on indicated that they carried wounded aboard.

After more manoeuvres and resupply on Saipan, the *Little* was sent into its first combat. They would take part in one of the most horrific battles of the War, on the island of Iwo Jima. On this small, five-mile by two-mile island, over 7,000 American and over 20,000 Japanese soldiers would die. The *Little* engaged in shore bombardment, landing escort duty, and laying down smokescreens to protect incoming troop and supply transport ships. From their offshore position, crew members witnessed an iconic moment, the famous flag-raising by Marines on Mount Suribachi. The *Little* performed well in its first combat engagement, earning its first battle star.

As the news of Iwo Jima reached the U.S., many Americans were aghast at the high number of U.S. casualties. Many were critical of the military for sacrificing so many there and wondered if this would be an indication of things to come, as the U.S. advanced closer and closer to the Japanese main islands.

Unfortunately, Iwo Jima turned out to be the "warm-up" for an even bigger engagement, the invasion of the Japanese island of Okinawa. The rocky island of Okinawa stood in the way of the U.S. planned invasion of the main Japanese islands, just 350 miles away. Initially, the *Little* was assigned to the diversionary force that would "fake" a landing on the opposite side of the island in an attempt to split the Japanese defenses. As the main invasion was well underway, the *Little* sailed to a position where they performed offshore "picket duty" with another destroyer, the U.S.S. *Aaron Ward* (DM-34), a Smith-class minelayer. Using on-board radar units, Frank and others in the radio crew watched for signs of Japanese aircraft.

They came under one direct daytime air attack and several night time raids. It was logged that they fired over 400 rounds of antiaircraft shells.

After a number of days manning their positions around the clock, the men were exhausted from lack of sleep. Tempers flared, and a few fistfights broke out on-board. The crew was given a few days to recover while the ship resupplied. Some played poker and shot craps. Sailors read and re-read letters they had received from home and wrote replies. They swapped stories of wives, kids, and girlfriends. Too soon, orders were received to head back to duty at Okinawa. The men mailed finished letters at the dock, stashed the letters they received in a safe place, and got ready to head back out to sea.

On May 3rd the *Little* arrived at Picket Station Number 10, accompanied again by the *Aaron Ward*. Four smaller craft, nicknamed the "pallbearers" accompanied them. At 14:15 four enemy planes appeared on the radar screen of the *Little*. The planes circled the ships from a distance but did not engage. At 18:15 a number of planes reappeared. The destroyers opened up with antiaircraft fire. At 18:43, as the sun was setting behind it, a single plane came through antiaircraft and machine gun fire and crashed into the port side of the *Little*. Undeterred, some crew members dealt with the damage, while the rest manned their battle stations. A second plane came into view and dove towards the ship. This time the *Little*'s gunners hit their mark, the plane was shot down, crash landing in the water just short of the ship. Another plane came in low, strafing the ship with machine gun fire. Thought to be a "Zeke," it came through the ship's fire and crashed into the starboard side between the aft and forward engine rooms. The *Little*'s gunners shot down another Japanese suicide plan that was making a near-vertical dive at the Aaron Ward. But another plane made a vertical dive at the Little and crashed into the aft torpedo mount. The three kami-kazes had likely coordinated the attack to damage the Little's midsection, in the area of its Number Two smokestack. The ship lost steering, power and communications. Its main deck leaned toward starboard taking on water. Captain Hall gave the final order to abandon ship at 18:51. The survivors floated in the open ocean in their life jackets. At 18:55 the U.S.S. Little's keel split in two and she sank midship sections first. The crew members watched in disbelief. As the sun was setting, the forward section of the ship stood straight up, then went down in what seemed like a final salute to her crew. In the final hours of daylight, the floating crew was subjected to strafing runs by Japanese Zeros. As Japanese planes circled the area, the familiar sound of a trio of Marine F-4U Corsairs was heard approaching. The Marine fighters jumped on the tails of the slower Japanese planes. When a couple of Japanese planes were hit and crashed into the water, the men cheered wildly. Since multiple air attacks were still in progress, the Navy waited a number of hours before starting a rescue operation. Hundreds of sailors were floating and treading water in the darkness. Many thought that they would be abandoned, possibly written-off as expendable. There was much worry for those in the water. The thought of sharks was terrifying. Would the Japanese fighters return in the morning to finish them off?

Deep in the night, off the coast of Okinawa, a bright searchlight cut through the darkness and shone on a group of floating men. The "pall-bearers" returned with another destroyer, the U.S.S. *Nicholson* (DD442). The survivors, wounded and dead were gathered into the rescue ships. Out of 200 crew members who started the day, 62 were killed and 27 were wounded. The *Little's* survivors were grateful that the Navy came to their rescue in the darkness. The rescue crews risked their lives on behalf of their fellow sailors. The crew members were not present when Okinawa finally fell to the Americans. It had been a costly battle and a harbinger of what an invasion of Japan's main islands would be like.

Mom spent time with her sister-in-law Josephine and their young son Raymond. They waited anxiously for news of Frank. The news from Okinawa was not encouraging, casualties were high, over 15,000 Amer-icans were killed, and over 40,000 wounded in fighting on land and sea. On Leyte, what MacArthur called a "mopping up" operation by the 8th Army would require months of tough fighting. And the 6th Army's casu-alties on the Philippine island of Luzon would reach over 8,000 killed and 30,000 wounded.

The Western Union courier stopped at the house again. This time his countenance did not signal bad news. A new telegram arrived from Frank himself, he had been rescued and was shipping back to Hawaii.

Mom continued to live with her parent's while Dad was overseas. She cared for their son, Edward, who would soon be turning two. He experienced a number of allergies from an early age. Mom and some of her friends worked a few hours at the packing house, wrapping bacon and other meats for shipment overseas. She said that it was actually fun, the women enjoyed some time out of the house, and as they talked and laughed, the time passed quickly. Plus, they were paid in cash. Mom put her extra funds into savings for use after the War. Both of Mom's brothers were now in the Navy, her younger brother Chester, recently joined, while older brother Frank recuperated in a Navy hospital in Hawaii.

America's ability to rapidly produce planes, tanks, ships, weapons, ammunition, and supplies for the troops has been noted by many as the key to Allied victory in World War II. The Germans and Japanese destroyed a significant number of Allied planes, tanks, and ships in the first part of the War. But as U.S. wartime production cranked up, early losses were more than made up for. The U.S. vastly outpaced their enemies in wartime production. Japan's and Germany's supply lines were gradually being cut off and their factories bombed. On the home front, posters of "Rosy the Riveter" rallied women to fill the gaps in factory jobs. African Americans and recent immigrants also entered the factories in record numbers. The War revitalized American industry. Automobile factories were converted to tank and jeep production. Women's clothing factories instead produced uniforms and parachutes. War bonds and stamps were sold to the public to fund the war effort. American citizens did without many items, as vital commodities and products were rationed, including gas, rubber, bacon, butter, and sugar. Other vital unsung contributors to victory were those who transported the massive amount of cargo along the vast supply chain. This included railroaders, truckers, dock workers, and the Merchant Marine who carried shipments across the oceans under threat of attack.

In Japan, women were also brought into the factory to fill the labor gap, as most men were conscripted into the military. One particular industry that Japanese women staffed was in the production of silk parachutes. Japanese paratroopers, the elite "Shinpei" or "soldier gods of the sky", were glorified in the culture. A spiritual connection was made between the women making the soldier's parachute and those that would use them in battle. This imagery inspired patriotic sentiment. A Japanese factory girl wrote a poem that reflects this:

> We young maidens brace up our frail bodies for the country.
> Today we will stitch the white silk again.
> While we make parachutes our mind is pure,
> strong, and calm just like the white silk.
> When we are stitching with our sewing machines our mind
> is just like that of a soldier carrying his gun to the battlefield.

Japan's government and propaganda machine seized upon this connection to motivate women to join the workforce and to inspire the men to fight on.

The Road to Surrender

I often watched the evening news with Dad. He related most to Walter Cronkite, known as the "most trusted man in America." Dad admired him as a "straight-shooter." In the late 60s and early 70s, the Vietnam War was usually the top story. Many had family members and friends who were serving overseas in the conflict. Vietnam was called the "living room war," since it was the first-time footage of a war of this level appeared mostly uncensored, on our living room TV sets. Cronkite went on location in 1968 following the Tet Offensive. His report of the situation was not complementary of U.S. leadership there. Dad agreed. He felt a great deal of empathy for the U.S. troops there. He'd often comment with disgust, "Mac and Krueger would never have left us like that." MacArthur and Krueger, much like Grant and Sherman of the Civil War, were associated with the principal of "total war." The Japanese were intent on fighting to the death, and the U.S. generals were eager to grant their wish in the most efficient way possible. Dad described the efforts in the Philippines to isolate the Japanese troops, cut off their supplies, shell them with artillery, and drive them out of their underground fortifications, as the way they were eventually defeated on the islands.

I didn't realize until later in life how the images of Vietnam must have triggered memories of the Philippines for Dad. The palm trees, jungles, rice paddies, bombings, the flames from napalm rising over the jungle hillsides, and the civilian population caught in the middle of fighting, must have looked very familiar. The beautiful scenery that would occasionally erupt into chaos. Napalm was first used by the U.S. in 1944. During fighting for the Philippines, Iwo Jima, and Okinawa, napalm was delivered in bombs and flame throwers to eliminate Japanese troops dug-into pillboxes and caves. But the images shown on TV of its devastating impact in Vietnam were instrumental in turning public opinion against its use.

Dad was not convinced of the U.S. government's premise of the War, portrayed as a fight against the spread of communism. He would sarcastically quip, "Want to know how to get the economy going? Start a war."

Dad admired MacArthur's planning skills and daring moves that often caught the Japanese by surprise. Even more important to Dad, was the perceived commitment MacArthur had for his troops. Dad always talked about how the General went "all-out" in efforts to defeat the enemy. Mac committed a superior number of troops to the major battles. He always attempted to provide more than enough supplies and support. Always on the offensive, he expertly applied the "island hopping" strategy. The strategy would bypass certain Japanese strongholds, and instead establish air bases on some key islands. Then bombers, Navy ships and submarines would cut off the supply lanes of the more heavily-defended islands, leaving the enemy troops there to "wither on the vine." This strategy was applied in the defeat of the Japanese on the Philippines. While MacArthur may have claimed that he invented the "island hopping" strategy, it actually came from the U.S. Navy. But MacArthur definitely perfected its application.

Dad talked about meetings he attended led by MacArthur. Mac would remove his hat at the conference table. Dad said that in the tropical humidity of the Philippines, MacArthur's comb-over would slowly start to rise off the side of his head. The officers, although wanting to bust out laughing at the sight, didn't dare. But it was a funny topic to talk about later over beers at the Officer's Club.

The fighting on Saipan, Iwo Jima, and Okinawa provided a sobering preview of what an invasion of the main Japanese islands would be like. The street-by-street fighting in Manila also indicated the potential impact on civilians. Over 100,000 civilian residents of Manila were killed in the fighting, many by the fires that ensued. Additionally, a large number were executed by the retreating Japanese. Many U.S. officials speculated about how high the casualty rates would be if the U.S. invaded the main Japanese islands. MacArthur downplayed the expected U.S. casualties, while many other estimates were shocking to the public. A study commissioned by Secretary of War Henry Stimson estimated 1.7 to 4 million American casualties, including 400,000 to 800,000 fatalities, and five to ten million Japanese fatalities.

Operation Downfall was the proposed Allied plan for the invasion of the Japanese homeland. It was planned to begin in November of 1945. Operation Downfall was divided into two parts: Operation Olympic and Operation Coronet. Operation Olympic would capture the southern part of the Japanese island, Kyūshū. The recently captured island of Okinawa would be used as the staging area. Then in early 1946 Operation Coronet would invade the Kantō Plain, near Tokyo, on the main Japanese island of Honshu. Airbases on Kyūshū captured in Operation Olympic would provide land-based air support for Operation Coronet. Operation Downfall was planned to be the largest amphibious operation in history, surpassing the Normandy Invasion in Europe and the Philippines Invasion.

The Japanese defense plan was called Operation Ketsugō. Japan's geog-raphy made it easy to predict where the U.S. would land, and the Japanese planned accordingly. They would make an all-out defense of Kyūshū, and leave little in reserve for any subsequent operations.

Japanese casualty predictions were also very high. It was expected that Japanese civilians would assist the military in resisting the invasion, by all means necessary, in accordance with the Emperor's demands. Japanese civilians, including the elderly, women and children underwent defensive training with crude weapons, such as sharpened bamboo sticks.

A seemingly harmless piece of pottery was a unique weapon produced to assist the Japanese citizenry in killing American invaders. Metal shortages in Japan limited the production of conventional grenades. But a number of craftspeople were available that were skilled in creating traditional ceramics in small operations across the islands. These people were commissioned to develop a ceramic hand grenade using their traditional techniques and small kilns. Spherical and about the size of a baseball, it was intended to be easily thrown by hand. Called the Type 4 grenade or "last-ditch grenade," it had a fragmentation body made of terra cotta or porcelain. It had a bottle neck with a rubber cover and a simple fuse attached to a detonator inside. Some were plain, porcelain white. Others were glazed in various colors by the artisans. A large number of these were produced and distributed to the public. Before the U.S. occupation following the surrender, much of the inventory, most without explosives, was dumped in the sea or buried by the Japanese. Stashes of these grenades have been found to this day throughout the Japanese home islands, some near former military installations.

As the 44th prepared for the planned invasion of Japan, there was very little hope of going home anytime soon. It was determined that the 44th would remain on Leyte staging for the invasion of Japan. Dad said that it was strange, that after years in the tropics, they were now gearing up with winter clothing. In the invasion planning, the 44th was to land on the main island of Honshu, near Yokohama, not far from Tokyo. They would be positioned in areas that would be cold with snowfall in the upcoming winter months.

The War Ends

My Aunt Annie, Dad's sister, lived in Independence, MO. She and my Uncle Ray owned a building a block or two from the Main Street Square. They operated a laundromat on the main floor, lived in an apartment above, and leased an adjacent apartment. Sometimes the family would meet there for dinner on Saturday and play cards late into the night. My older male cousins had a massive electric train set that took up a whole room. It was complete with towns, farms, mountains, tunnels, and bridges. Uncle Ray was a retired railroad engineer and conductor. His collection of railroad memorabilia covered the walls. Sometimes I'd spend the night, and Mom and Dad would pick me up in the morning on their way to Sunday Mass at the Cathedral downtown. Former President Harry S. Truman lived only a few blocks away. After retiring to Independence, he was known for taking morning walks. Sometimes he'd be accompanied by his own private security guard. It wasn't until 1965 that former presidents received lifetime Secret Service protection.

One Sunday morning in the early '60s, when Mom and Dad picked me up, we noticed the former President, walking alone and seemingly deep in thought. He was dressed in a dark suit and tie, with his cane and trademark hat, a light-grey, custom-made Stetson fedora. Truman dressed with style. Prior to getting into politics, he had owned a haberdashery in Kansas City. As I sat in the backseat of our '62 Ford Galaxie, I gazed out the window at him. There he was, the 33rd U.S. President, the man who ended World War II and ushered in the atomic age. He gave me a nod, looked down to check the position of his cane, and continued walking. I turned around to look at him through the rear window as we drove away. A man who changed history, he looked like any other citizen, out for a stroll on this quiet Sunday morning.

That night, as we watched the Russian satellite Sputnik streak across the dark clear sky over the patio, I asked Dad "Do you think that they'll ever be a nuclear war?" We recently had a close call in the Cuban Missile Crisis. My brother Ed, a Marine, was sent to Cuba.

I also asked Dad, "Do you think that it was right to use the bombs on Japan?" These were tough questions. He responded, as many veterans did, that he believed the bomb had saved his life. He described that the 44th had continued their preparation for the invasion of Japan. Then, one day, all of the planning came to a standstill. All he knew, was that now he would be going home.

The first atomic bomb used in warfare was dropped on Hiroshima at 8:15 a.m. on August 6, 1945. Most of the city was destroyed and approximately 80,000 people were killed in an instant. By the end of the year 90,000 to 166,000 had died as a result of the blast and its effects. A second bomb was intended for the city of Kokura, but due to cloud cover, an alternate target was selected, the city of Nagasaki. As the last atomic bomb used in warfare, over 40,000 Japanese citizens were killed instantly. Many more would die from wounds and the effects of radiation.

The bomb became the new threat to the existence of civilization. "There would never be another war like World War II," some members of the 44th quipped after the announcement of the bombing. Some sarcastically commented, that "they just took all the fun out of it."

The surrender of Imperial Japan was announced by Japanese Emperor Hirohito on August 15, 1945. Historians today speculate whether the atomic bombs made the difference, or the fact that the Russian Army entered the war against Japan. Given the choice, the U.S. terms of surrender likely seemed to be the better option. Russia, a former enemy of Japan, was brutal in their final push against Germany. The U.S. offered a way for the Japanese Emperor to survive, as a figurehead to the people, with no political power.

The directive came to the 44th and all troops on the Philippines, "Do not fire unless fired upon." There was a jubilant celebration. The 44th had a front row seat on the beach that evening as ships in Tacloban's harbor fired shells and tracers to celebrate the end of the War. "Golden Gate in '48" turned to "Back Alive in '45." But there was still an uneasiness within the ranks. There were many uncertainties. Would the Japanese actually stop fighting? Will the Japanese military leaders in Tokyo accept the terms or lead a revolt? What would occupation of Japan be like? The 44th had taken in many casualties from the brutal fighting on Luzon. There were concerns regarding how long would it take to transport the wounded back to the States. Even though the fighting in the Philippines mostly ceased, there was still much work in treating and evacuating the wounded. This now included those liberated from military and civilian prison camps.

The Santo Tomas Internment Camp, also known as the Manila Intern-ment Camp, was the largest of several camps in the Philippines in which the Japanese interned enemy civilians, mostly Americans, during the War. The campus of the University of Santo Tomas in Manila was utilized for the camp, which housed more than 3,000 internees from January 1942 until February 1945.

The internees were diverse: business executives, mining engineers, bankers, plantation owners, seamen, shoemakers, waiters, beachcombers, prostitutes, retired soldiers from the Spanish–American War, missionaries, and others. Some came into the camp with their pockets full of money and numerous friends on the outside; others had only the clothes on their backs. During the War, a total of about 7,000 people were resident in Santo Tomas. There was a regular flow of people in and out of the camp, as some missionaries, elderly, and sick people were initially allowed to live outside the camp and more than 2,000 were transferred to Los Baños internment camp. About 150 internees were repatriated to their home countries as part of prisoner exchange agreements between Japan and the United States and the United Kingdom. Most internees, however, served a full 37 months in captivity.

At first, most internees believed that their imprisonment would only last a few weeks, anticipating that the United States would quickly defeat Japan. As news of the surrender of American forces at Bataan and Corregidor reached the camp, the internees settled in for a long stay. Conditions rapidly deteriorated during the War and by the time the camp was liberated by the U.S. Army, many of the internees were near death from starvation and disease.

The Santo Tomas internees began to hear news of American military action near the Philippines in August 1944. Clandestine radios in the camp enabled them to keep track of major events. On September 21st the first American air raid struck the Manila area. As American forces invaded the Philippine Island of Leyte on October 20, 1944, and later advanced on Japanese forces occupying other islands in the country, U.S. planes bombed Manila on a daily basis.

On December 23, 1944, the Japanese arrested four camp leaders for unknown reasons. Speculation was that they were arrested because they were in contact with Filipino soldiers and guerrilla resistance forces. On January 5th, the four men were removed from the camp by Japanese military police. Their fate was unknown until February when their bodies were found. They had been executed. The U.S. rushed to liberate the prisoner of war and internee camps in the Philippines due to a common belief that the Japanese would massacre all their prisoners, military and civilian.

A small American force pushed rapidly forward and, on February 3, 1945, at 8:40 p.m., internees heard the sound of tanks, grenades, and rifle fire near the front wall of Santo Tomas. Five American tanks from the 44th Tank Battalion broke through the fence of the compound. The Japanese soldiers took refuge in the large, three-story Education Building, taking 200 internees hostage, including internee leader Earl Carroll, and interpreter Ernest Stanley. Carroll and Stanley were ordered to accompany several Japanese soldiers to a meeting with American forces to negotiate a safe passage for the Japanese out of Santo Tomas in exchange for a release of their 200 hostages. During the meeting between the Americans, Filipinos and Japanese, a Japanese officer named Abiko reached into a pouch on his back, apparently for a hand grenade, and an American soldier shot and wounded him. Abiko was especially hated by the internees. He was carried away by a mob of enraged internees, kicked and slashed with knives, and thrown out of a hospital bed onto the floor. He died a few hours later.

The evacuation of the internees began on February 11th. Sixty-four U.S. Army and Navy nurses interned in Santo Tomas were the first to leave that day and board airplanes for the United States. Flights and ships to the United States for most internees began on February 22. Although food became adequate with the arrival of American soldiers, life continued to be difficult.

The lingering effects of near-starvation for so many months saw 48 people die in the camp in February, the highest death total for any month.

Several army nurses were interned with civilians at Santo Tomas. Mrs. Nancy Belle Norton, the "Angel of the Philippines" was awarded the Medal of Freedom for her actions helping fellow POW during the war.

Following V-J Day, Colonel Weston was ordered to send some of the 44th's doctors and nurses to assist with the treatment and evacuation of civilians at Santo Tomas. Doctors Pyre, Tausend, Midelfart, and Daniels, and a small group of nurses, were sent by plane to Manila. Dr. Pyre described the scene upon landing at Clark Field, "hundreds of disabled Jap planes, riddled by machine gun fire, were lined up on the field." He then described the conditions at Santo Tomas, which was not quite free from Japanese attack, "(the Japs) were throwing shells. These shells came from across the river. One shell, and I will never forget the sound of an incoming missile, burst through the window of a large 'hospital ward' where I was attending the recently liberated internees. The bulk of the charge demolished an Army Chaplain at the other side of the big room. His brains and scalp smeared the ceiling." Some of the internees from Santo Tomas and from liberated Japanese prison camps on Luzon were evacuated to the 44th General Hospital at Tacloban. Pyre and the other doctors received the Bronze Star for their work at Santo Tomas. The Bronze Star is a U.S. decoration awarded for heroic or meritorious service in a combat zone.

The Japanese surrender papers were formally signed on September 2, 1945, aboard the battleship *Missouri* in Tokyo Harbor. The Japanese surrender brought the most devastating war in history to a close. The most powerful nations in the world had clashed, now only the hope of a better future lay ahead for the generation that lived through it. Maybe General Douglas MacArthur put it best in his speech at the Japanese surrender ceremony, 75 years ago,

> *The issues involving divergent ideals and ideologies have been determined on the battlefields of the world, and hence are not for our discussion or debate.*

*Nor is it for us here to meet, representing as we do a majority of
the peoples of the earth, in a spirit of distrust, malice, or hatred.*

*But rather it is for us, both victors and vanquished, to rise to that
higher dignity which alone befits the sacred purposes we are about to
serve, committing all of our peoples unreservedly to faithful compliance
with the undertakings they are here formally to assume.*

*It is my earnest hope, and indeed the hope of all mankind, that from
this solemn occasion a better world shall emerge out of the blood and
carnage of the past -- a world founded upon faith and understanding,
a world dedicated to the dignity of man and the fulfillment of his most
cherished wish for freedom, tolerance, and justice.*

MacArthur wanted officers over six feet tall to be part of his occupying
force, as he wanted them to tower over the Japanese people. Dad, now a
captain, and a lanky 6' 2", was asked if he would accept a position in the
occupation force. A promotion to major would be provided if he accepted.
Dad sent a telegram to Mom, optimistically asking her if she would travel
to Japan and join him. He described that they would live in a mansion,
that she would have servants, an official car, and driver. He also said that
they would have a security detail, so she wouldn't have to worry about
their safety. Mom sent a reply that flatly refused. She stated that little
Eddie was having issues with allergies and that she was concerned that
the Japanese might still retaliate against the occupying Americans. She
thought it was best if he came home. Dad reluctantly agreed. With his
rank, recognitions, and time overseas, Dad's ASR score was high enough
that he could ship out soon.

On November 13, 1945, Dad and other members of the 44th boarded
a ship for San Francisco. They passed back under the Golden Gate Bridge
on November 27th. Dad, having just spent two years in the tropics, said
that he had never felt colder than in the damp San Francisco Bay. Once
onshore, he and many of the other officers obtained the heaviest
sweaters they could find. Dad wired Mom and said that he looked
forward to a peaceful Christmas at home. He brought his son Eddie,
now just over

two years old, a stuffed Koala Bear from Australia, made of wool. He also presented his Dad with an M1 carbine, the one he used on Leyte. Dad always described how little Eddie didn't know who he was and was very upset that this "stranger" was now staying with him and his mother. As with all veterans, recovering from the War and getting used to civilian life again would take some time.

Fortunately, the occupation of Japan and post-war relations were peaceful. The Emperor was allowed to stay in power, but mostly as a figurehead to the people. The U.S. poured aid into rebuilding the devastated Japanese economy. Military figures faced charges in military courts.

Japanese General Yamashita was tried and found guilty of war crimes committed by troops under his command during the Japanese defense of the Philippines in 1944. In a controversial trial, Yamashita was found guilty of his troops' atrocities even though there was no evidence that he approved or even knew of them. Yamashita was sentenced to death and executed by hanging on February 23, 1946. The ruling against Yamashita, holding the commander responsible for subordinates' war crimes as long as the commander did not attempt to discover and stop them from occurring, came to be known as the "Yamashita Standard." In recent times, Yamashita has become famous again, as he has been linked with the confiscation and hiding of gold acquired in Japan's conquests. It's alleged that he transported vast amounts of gold from conquered nations to the Philippines, where he hid it in multiple locations across the islands. It's also alleged that he sealed off caves with explosives where the gold was stashed, sometimes with laborers who transported it inside. A number of books and documentaries on the "Yamashita's Gold" mystery are currently available.

General Suzuki was subsequently killed in the Philippines by a U.S. air attack in April of 1945. General Shiro Makino, leader of the 16th Infantry Division, took his own life on August 10, 1945, after the battle for the Philippines was lost. Only a few hundred men from his original division of over 10,000 stationed on Leyte survived the War.

General Douglas MacArthur received the Medal of Honor for his service in the Philippines campaign, which made him and his father, Arthur

MacArthur Jr., the first father and son duo to be awarded the medal. After disagreements with President Truman over the Korean War, MacArthur was contentiously removed from command on April 11, 1951. He then retired from the Army and held various private positions. In the 1960s he advised both President Kennedy and Johnson against involvement in Vietnam. MacArthur passed away on April 5, 1964, at the age of 84. His controversial legacy is the still the subject of much debate. But for a large number of the troops he led, there was a deep connection and reverence for the man. There were only a few times I remember seeing Dad cry. One of those times was on the announcement of MacArthur's death.

The members of the 44th received The Meritorious Unit Commendation (MUC, pronounced muck) for their performance at Burauen in the "Battle of Buffalo Wallow." The MUC is a mid-level unit award of the United States Armed Forces. The U.S. Army awards units the Army MUC for exceptionally meritorious conduct in performance of outstanding achievement or service in combat or non-combat duties. Troops nicknamed it "the golden toilet seat" for its appearance. The 44th also received the Philippine Liberation Medal, a military award of the Republic of the Philippines which was created by an order of Commonwealth Army of the Philippines Headquarters on December 20, 1944.

On the 75th anniversary of the end of World War II, I looked through Dad's photo album. The images of the people and places enriched and personalized Dad's stories for me. There are photos of the dedicated men and women of the 44th, smiling in the midst of the hardships of war. Photos of the proud New Guinea natives, their primitive appearance in sharp contrast to the uniformed U.S. Army officers. The faces of the Filipinos, their fighting spirit prevailing, their freedom and smiles returning. The solemn looks on the faces of the orphaned children. Not knowing anything but war, they seemed much older than their years. I look at photos of defiant Filipino guerrillas. One Filipino man raising a bolo knife, had the fierce look of determination on his face. His young son is by his side, with the same look on his face. A small bolo knife is attached to his belt. He grasps the handle, as if ready to use it. He was born into war. I can't imagine if that was my son. I greatly respect and admire the determination of the Filipinos through the War. Through these photos and the stories of their experiences, I feel that I know them better.

My thoughts are more divided when I look upon Dad's photos of the Japanese. Their brutality to the Filipino men, women, and children was inexcusable. Their treatment of prisoners of war was worthy of the war crimes they were prosecuted for. They killed over 3,500 U.S. servicemen on Leyte; my Father, fortunately, was not counted among those. But the Japanese soldier was committed to fighting to the death, in spite of his hardships. Of the 49,000 Japanese killed on Leyte, it's estimated that 80% died of starvation or disease. Very few Japanese surrendered. In surrendering, they would face shame from their families and country.

I look at the photos of the dozen or so Japanese prisoners held in a stockade that Dad managed. A few of them look defeated, but most others look relieved. Their struggle was over; they would at least live beyond the War. I look at the photos of the dead Japanese soldiers that I used to sneak to see when I was very young. They lie in the buffalo wallow where they fell, face down in the mud, some with limbs torn off. Two wore the uniform of a paratrooper, one of the infantry. The infantry soldier, his body twisted to one side, died clutching his sword to the end. These men fought and died courageously on Leyte. The samurai warrior spirit lived within them. When younger, I would look at the photos with a shallow curiosity. Now I know more of their story. I can now at least appreciate who they were. And in spite of their brutality, I nonetheless can't help but respect their tenacity, commitment and endurance to their cause.

I hold the artifacts that Dad returned home with. The Japanese sword that was clutched by the infantryman in the wallow, the bolo knife that the defiant Filipino guerrilla used to protect his family, and the caduceus emblem worn by my Dad and the members of the 44th General Hospital. I feel that I got to know something significant about those who possessed these items. I understand more how these diverse people came together in the world's largest conflict. I respect their courage and commitment in fighting to uphold their principles. I echo General MacArthur's sentiment, in the hope that freedom, tolerance, and justice will prevail.

The veterans of the 44th never complained about their circumstances. But, later in their lives, some veterans of the 44th were disturbed by the fact that official Army records did not account for their service on Leyte. Many believed that there was an intentional cover-up to protect high-level commanders from the decisions that were made. Seventy-five years after the end of the War, does it matter? Time has erased any consequences to the commanders who are now long gone. Maybe this was a reflection of a different time, when the message was controlled by commanders such as MacArthur. During the period of World War II, news events described by military leadership were bent towards self-glorification. Pride may have led to the suppression of accurate information. The news rarely criticized actions or provided a less than "rosy" analysis of those in command. The chaos of war, particularly in the era where technology and communications were less than reliable, inevitably created issues. The Korean and Vietnam conflicts opened up a more critical assessment of military mistakes. We risk losing the real truth from the "Greatest Generation" that participated in the largest conflict in history. Only by surfacing the stories of those who were actually there, we can get a true perspective of events, the good, the bad, and the ugly.

The stories of the veterans of the 44th General Hospital persist to this day, as if broadcast through the ether. Curiosity led me to validate the stories I heard my Dad tell on the patio where we sat on those warm summer nights. My research led to a number of other veterans of the 44th who likewise shared their experiences. Their unit, like many others, had formed a unique, common bond through their service. I think that they realized, in the twilight of their lives, that they weren't "just doing their jobs," they were indeed extraordinary in the part that they played. Healers, defenders of other's freedom, dedicated citizen soldiers who had the courage to fight when they had to. I believe that General MacArthur was willing to sacrifice the service personnel remaining at Burauen in pursuit of his timeline on Leyte. Army Headquarters dismissed the intelligence reports that forewarned of a planned Japanese counter-attack on the Burauen airfields. A senior group of highly-skilled civilian physicians were not even evacuated. The members of the 44th were left to defend themselves and their patients. It appears that even with the availability of intelligence, its still up to the military leaders to decide how to act upon it, given their overall objectives.

Leyte was the major turning point in the War of the Pacific, and the 44th played a key part in it. I'm proud to tell their story. I believe that it has a unique place in the many stories of courage and dedication of our World War II veterans. On the faded black and white photos taken by Walter Teague and my Dad, I see the smiles of the Filipino men, women and children. I think that I realize why the 44th General Hospital was motivated to act with courage, and like the samurai, defend their principles to the end.

It's easy for me to remember the key dates of World War II. My brother was born on September 2nd, the date of the official end of World War II. I was born on May 8th, the date that Germany surrendered, Victory in Europe Day (V-E Day). And Dad passed away on August 15, 1989, at the age of 75, Victory over Japan Day (V-J Day).

The Buffalo Wallow Fight of 1874

The original Battle of Buffalo Wallow was fought in 1874 as part of the Red River War at a site twenty-two miles southeast of Canadian, TX, in the Texas Panhandle. The Red River War was a military campaign launched by the U.S. Army in 1874 to displace the Comanche, Kiowa, Southern Cheyenne, and Arapaho Native American tribes from the Southern Plains, and force them to relocate at reservations in Indian Territory (now the state of Oklahoma). On September 10, 1874, Colonel Nelson A. Miles, whose command was running short of rations, sent two scouts, Billy Dixon and Amos Chapman, and four enlisted men, from his camp on McClellan Creek to check on the delay of a train of supply wagons. It happened that the supply wagons were under siege by Indians on the upper Washita River. The six men set out on the trail to Camp Supply through Indian Territory. On the morning of September 12th, as the scouts approached the Washita River, they suddenly found themselves surrounded by over 100 Comanche and Kiowa warriors, some of whom had come from the supply train attack. Since the warriors had burned off the prairie grass only days before, there was limited shelter on the treeless prairie. Dixon and his companions were forced to dismount and make a desperate stand on the ground with their saddles shielding them from the warrior's bullets, arrows, and spears. In the first few minutes of the attack, the enlisted man who held the horses was shot and killed. The horses stampeded, carrying with them the men's haversacks, ammunition, canteens, coats, and blankets. The native warriors, expert horsemen, circled the men as they fired on the run from the sides of their mounts. All of the cavalrymen were hit, some multiple times. Dixon's loose shirt had several holes, the bullets barely missing his body. He was then hit in the calf. Amos Chapman's left knee was shattered by a bullet. When the Indians stopped to reload their rifles, Dixon, taking advantage of the lull, scanned the prairie for a better defensive position. He spotted a buffalo wallow fifty or so yards away. He was able to lead all but two of the others to the shallow depression, about

179

ten feet in diameter and a foot deep. With their bare hands and butcher knives the men dug deeper into the sandy loam around the perimeter of the wallow. In the process, the men managed to keep their adversaries at bay and away from the other two enlisted men still out on the prairie in front of them.

As the fight progressed, Billy Dixon tried several times to reach the stranded men, but was forced back repeatedly by a hail of bullets and arrows. Amos Chapman had lived among the Indians for a time and was known to many of the warriors. They taunted him by shouting, "Amos, Amos, we got you now, Amos!" Then, early in the afternoon, Dixon made it to one of the men and carried him back amid gunfire to the relative safety of the wallow.

As the day wore on, the five men suffered from hunger, thirst, and numerous wounds. They continued to hold back the Indians, who were still trying to kill the lone man outside of the wallow. Late in the afternoon a thunderstorm brought much needed relief. The water in the wallow mixed with mud and their own blood, but the thirsty men drank it anyway. The rain caused a sharp drop in temperature as the evening arrived and the men were cold and suffering from shock. The men assumed that their stranded companion was dead. Taking advantage of a break in the fighting, Dixon went out to recover the dead man's weapons and ammunition. He was astonished to find the man still alive. He carried the soldier back to the wallow. The wounded soldier died from his wounds later that night.

At nightfall the Indians disappeared. The surviving men created crude beds for themselves and their wounded comrades out of tumbleweeds they had gathered and crushed. The night was miserably cold and wet. The following morning, September 13th, was clear and mild. To the group's astonishment, there were no Indians in sight. Dixon volunteered to go for help. As he made his way back to the trail, he headed back towards the fort. In the distance, he spotted a column of men on horseback and fired his gun to get their attention. It turned out to be four companies of the Eighth Cavalry from Fort Union, New Mexico, numbering over

200 men. It was the large contingent's appearance that had caused the Indians to withdraw from the siege of the wallow and the supply wagons.

Dixon led a cavalry column back to the site of the battle. The cavalry unit had no ambulance wagon and they were also low on supplies. The unit's surgeon accompanied Dixon in the lead. From the distance, the frightened men at the wallow mistook the approaching column for Indians and shot the horse from under one of the surgeon's assistants. The surgeon only briefly examined the wounded and starving men, and they were only given a meager amount of food and water. The cavalry moved on with their mission, promising to send help. Colonel Miles had not sent aid either. Since they had not returned with any news, he likely supposed that they had all been killed.

At midnight, however, aid finally arrived. The wounded and nearly dead men received some food and medical attention. The men who died were wrapped in Army blankets and buried in the wallow. The wounded were taken to Camp Supply for treatment at the camp's hospital. Amos Chapman's leg was subsequently amputated above the knee.

Colonel Miles censured the Eighth Cavalry Commander for his failure to render immediate assistance to the survivors. Miles recommended that the surviving men be awarded the Congressional Medal of Honor for bravery under adverse circumstances. The medals were given, personally by Miles, to Dixon and Chapman; a posthumous one was awarded to the man who held out by himself on the prairie during the attack.

The Buffalo Wallow Fight was widely publicized as a heroic engagement of the Red River War. On his deathbed Billy Dixon told his wife Olive the story of his life, and the Buffalo Wallow Fight, which she penned and was later published as *The Life of Billy Dixon*. Colonel Richard Irving Dodge wrote a conflicting account of the encounter in an 1882 book titled *Our Wild Indians*. In that he stated that no Indians were killed. Dixon and the other surviving members resented the account, which they said was gathered from second hand sources. In 1917 the medals of Dixon and Chapman were revoked by Congress on the technicality that they had served in the Army as *civilian* scouts. Dixon, however, refused to surrender what he felt he had courageously earned. His medal is now on display at the Panhandle-Plains Historical Museum in Canyon, TX, south of Amarillo. In 1925, a granite monument was erected on the Buffalo Wallow battle-field site. It bears the names of the six men who fought there. In 1989 an Army Board of Correction of Records reinstated the awards of Dixon and Chapman. There have only been six other civilians who have been awarded the Medal of Honor.

Sherman Miles, Nelson Miles's son, was also a general in the U.S. Army. He was Chief of the Military Intelligence Division in 1941, when the attack on Pearl Harbor occurred. He had been criticized after the War, along with higher ranking U.S. officials, for not heeding the intelligence gathered about a possible Japanese attack on the U.S. Naval fleet. Miles defended his actions and implied that other circumstances enabled the Japanese surprise attack. Over 75 years later, controversy still abounds as to whether those stationed at Pearl Harbor were deemed expendable, in order to accomplish the larger objective of getting America involved in the War.

The phrase "rich man's war, poor man's fight" was a rallying cry that originated during the Civil War. Southern plantation owners sent thousands of poorer citizens to their deaths to defend their economic way of life and the institution of slavery. Wealthy northern families could donate $300 or find a substitute to get their sons out of the draft. These have been common themes throughout time, where the individuals who fight the wars are sacrificed on the battlefield for the greater objectives of the nation or those in power.

It seems that many who serve in wars, from those who perished at Pearl Harbor, to the support troops at the airfields of Burauen, to the sailors floating in the sea off Okinawa, to the Japanese who made their last stand on Leyte, to those who came later to fight in Cuba and Vietnam, and to the six men at the Buffalo Wallow Fight in 1874; that sometimes they are left to die, "face down in the mud." Then there are those in the legions that fight until the end, maybe not to achieve the objectives of the nation, but rather to defend their ideals and the lives of their brothers who fight beside them. I particularly appreciate the medical staff who work to heal and preserve life during the world's conflicts. Unique among them, are the veterans of the 44th General Hospital, who proved that they could be both warriors and healers. I may not understand the complexities of war, but I do thank our World War II veterans for saving the world at a critical time in history, and preserving the freedom of people in America, the Philippines, and many other nations.

On a warm summer day, a lone eagle flies above Cheyenne Mountain, its wings outstretched and still, it drifts effortlessly in the thermals above the forest below. In Colorado Springs, a 100-year-old Kiowa Medicine Man passes away. Another warrior and healer fades away.

The Battle of Leyte is Not Yet Over

Chieko Takemi published a book in 1999 called *The Hidden Battle of Leyte: A Picture Diary of a Girl Taken by the Japanese Military*. It tells the tragic story of Remedios Felias, a 14-year-old Filipina taken by the Japanese into sexual slavery during the occupation of Leyte. Remedios grew up in Esperanza Village, Burauen, Leyte. She was present when the Japanese invaded the Philippines. She was captured as she and her family tried to flee Burauen as the Japanese moved in to establish their air bases in Central Leyte. She became a Japanese "Comfort Woman," held within a Comfort House near the Japanese base. She was brutally beaten and raped repeatedly over two years until the U.S. Army invaded and liberated the Burauen area. After having spent five days in a Japanese bomb shelter without food, she came out of the bunker after her captors either fled or were killed. Thought to be a Japanese soldier, she was almost shot by the Americans, until a Filipino guide recognized that she was a young girl. She was taken by Army jeep to the 44th General Hospital for immediate treatment. After rehabilitating, she was able to reunite with her family. She dealt with the effects of the abuse for the rest of her life. Much later in her life she was connected with other Comfort Women in Manila. There, she and other women drew colored pencil sketches of their experiences. Chieko Takemi, a Japanese female journalist, was greatly moved by her story and published Remedios's sketches and descriptions of events, translated from Tagalog to English and Japanese. This very moving and beautifully illustrated book is available from Amazon. Chieko worked diligently on the efforts to gain restitution for the Comfort Women from the Japanese government, and also to help raise funds to restore the agricultural life of the people of Leyte. She initiated a program that would use the funds to purchase carabao for the farmers, who lost most of their stock during World War II.

In honor of the veterans of the 44th General Hospital, 10% of the proceeds of the "Buffalo Wallow" book will be donated to the cause of women impacted by war and sexual slavery. To donate to the rights of women across the world, please consider the following organizations:

- Lila Pilipina, an organization that works for victims of sexual abuse during military intervention and acts of aggression. They are advocates for those impacted by military sexual crimes. They've provided assistance and advocacy for the victims of sexual slavery ("comfort women") imposed by the Japanese Army during World War II. Please see the website: https://movementforthelolas.wordpress.com/background/.
- Conscious by Kali, an organization founded by Kali Basi to assist women, many of them very young, impacted by human trafficking. My wife and I met Kali in Madison, WI, while doing research on this book. Her website is: https://www.consciousbykali.com/.
- Reasons Rescue Ranch, a non-profit ranch in the Missouri Ozarks operated by my sister, Rita, to provide a home for rescued horses and other animals. Rita pairs up young women with horses that they care for and learn to ride: https://reasonsrescueranch.org/about-us.

Epilogue

In July of 2014 our family met for a camping trip at Montauk State Park, a hilly, forested area deep in the Missouri Ozarks. It had been one of Mom and Dad's favorite getaway places. The cold clear Current River starts there from an eternally bubbling spring. The stream winds through the hills of oak, pine, dogwood, and redbud. Trout thrive in the gently flowing waters and amidst the clumps of watercress that grow wild in the stream. We met here in a special remembrance of what would have been Dad's 100th birthday. Mom lived to the age of 92 and passed away in 2010. The family, including their children, grandchildren, and great-grandchildren reminisced about the time we spent there. It was at that time that I had started my research into Dad's War experience. In the years that followed, with his Service Record in-hand and repeated Google searches, I was able to connect the dots and link the historical context with his stories.

I was also extremely fortunate to become familiar with others that served with the 44th. Many of the doctors and nurses continued to serve in the Army Medical Unit until their retirement. Others returned to civilian life and had very successful private practices.

Many of the 44th's veterans kept in touch after the War. After serving with such a talented group of physicians during the War, Dad always had a number of "war buddies" that he could call on for second opinions. In his later years, Dad had bouts with heart disease, stroke, and prostate cancer. Those phone calls would always include talk of family and baseball. I don't recall them ever talking of the War.

The Teagues kept a large amount of correspondence with fellow veterans after their retirement. They attempted to form a reunion for the 50th anniversary of the Japanese surrender and travel back to Australia and the Philippines. But by that time in 1995, many of the veterans had health issues that precluded them from going. Some of them wrote letters in reply, recounting their stories of the War and the Battle of Buffalo Wallow. These were priceless. Walter and Eda made the trip back to Australia and to Leyte.

The articles they wrote and their Library of Congress interviews were also priceless. The Teague's spoke at secondary schools about the experiences in serving in the military medical service. I'm sure that they've inspired many in the generations that followed them. The Teague's materials are archived at the Wisconsin Veteran's Museum in Madison, WI. As this was where the 44th's medical staff originated, it is a fitting place to house their memories.

The veterans of the 44th have a unique story to tell. My Dad's and other eyewitness accounts tell a story of courage, dedication, and resolve. The veteran's stories didn't exaggerate what they did. After what they experienced, I think that they were humbled by the fact that they made it back alive. While so many others they served did not. In the way they described it, they were "just doing their job." As it turned out, it was what Dad didn't tell me about his experience that said the most. Later, I realized what their "job" included.

The U.S. Congress asked General Douglas MacArthur to address a joint session on April 19, 1951, as he ended his long military career. Although President Truman had relieved him of his duties during the Korean War, he was still recognized as one of the nation's greatest living military leaders. His speech is known for its famous line, in which he quoted an old army ballad:

I am closing my 52 years of military service. When I joined the Army, even before the turn of the century, it was the fulfillment of all of my boyish hopes and dreams. The world has turned over many times since I took the oath on the plain at West Point, and the hopes and dreams have long since vanished, but I still remember the refrain of one of the most popular barrack ballads of that day which proclaimed most proudly that "old soldiers never die; they just fade away." And like the old soldier of that ballad, I now close my military career and just fade away, an old soldier who tried to do his duty as God gave him the light to see that duty. Good Bye.

Although many of the citizen soldiers of the 44th General Hospital lived into old age, their lives and accomplishments, like those of their leaders, have "faded way" into the history of World War II. I hope that this book is a fitting memorial to them.

References

Books and Articles

Borneman, Walter R. *MacArthur at War: World War II in the Pacific.* Walter R. Borneman, 2016.

Cannon, M. Hanlon. *Leyte: The Return to the Philippines.* Washington, DC: Office of the Chief of Military History, 1954.

Charles River Editors. *The Philippines Campaigns of World War II: The History of the Japanese Invasion in 1941–1942 and the Allied Liberation in 1944–1945.* Charles River Editors, 2014

Chun, Clayton. *Leyte 1944: Return to the Philippines.* Oxford: Osprey Publishing, 2015.

Condon-Rall, Mary Ellen and Albert E. Cowdrey. *The Medical Department: Medical Service in the War Against Japan.* Washington, DC: U.S. Center of Military History, 1997.

Costello, John. *The Pacific War, 1941–1945.* New York, NY: Harper Perennial, 1981.

Dencker, Donald O. *Love Company: Infantry Combat Against the Japanese, World War II*, Manhattan, KS: Sunflower University Press, 2002.

Dixon, Olive K. *The Buffalo Wallow Fight: Extract from the Life of Billy Dixon.* Dallas, TX: P. L. Turner Company, 1935.

Duffy, James P. *War at the End of the World: Douglas MacArthur and the Forgotten Fight for New Guinea, 1942–1945.* James P. Duffy, 2016.

Dunant, Henry. *A Memory of Solferino, 1862*, translated into English by the American Red Cross.

Felias, Remedios and Chieko Takemi. *The Hidden Battle of Leyte: The Picture Diary of a Girl Taken by the Japanese Military.* Tokyo: Bucung Bucong Publishing, 1999.

Heinrichs, Waldo and Marc Gallicchio. *Implacable Foes: War in the Pacific: 1944-1945.* New York, NY: Oxford University Press, 2017.

Henson, Maria Rosa. *Comfort Woman: Slave of Destiny.* Manila, Philippines: Philippine Center for Investigative Journalism, 1996.

Holzimmer, Kevin C. *General Walter Krueger: Unsung Hero of the Pacific War.* Lawrence, KS: University Press of Kansas, 2007.

Kirk, Norman T. "That They May Live!—A Pictorial Story of the Army Medical Corps, Major General Norman T. Kirk, Surgeon General of the Army." *Collier's* (July 21, 1945): 15.

Krueger, General Walter. *From Down Under to Nippon: The Story of Sixth Army in World War II.* Washington, DC: Combat Forces Press, 1953.

Oki Tsuruma, Kōgyō. *Rakkasan o Tsukuru Kokoro (The Spirit of Making Parachutes).* Tokyo: Fujikura Kōgyō KK, 1943.

Prados, John. *Storm over Leyte: The Philippine Invasion and the Destruction of the Japanese Navy.* John Prados, 2016.

Prefer, Nathan N. *Leyte 1944: The Soldier's Battle.* Havertown, PA and Oxford: Casemate, 2012.

Pullman, Sally Hitchcock. *Letters Home: Memoirs of One Army Nurse in the Southwest Pacific in World War II.* AuthorHouse, 2004.

Rottman, G. and A. Takizawa. *Japanese Paratroop Forces of World War II.* Oxford: Osprey Publishing, 2005.

Salecker, Gene Eric. *Blossoming Silk Against the Rising Sun: U.S. and Japanese Paratroopers at War in the Pacific in WWII*. Mechanicsburg, PA: Stackpole Books, 2010.

Serling, Anne. *As I Knew Him: My Dad Rod Serling*. New York, NY: Kensington Publishing Corp., 2013.

Sharpe, Dr. George. *Brothers Beyond Blood: A Battalion Surgeon in the South Pacific*. Austin, TX: Diamond Books, 1989.

Special Service Division Services of Supply U.S. Army. *Instructions for American Servicemen in Australia 1942*. U.S. Army, 1942.

Spector, Ronald H. *Eagle Against the Sun: The American War with Japan*. New York, NY: Random House, 1985.

The State Medical Society of Wisconsin. *War without Guns: A Record of Service of Wisconsin Physicians in World War II*. The State Medical Society of Wisconsin, 1949.

Xu, Klytie, Stacey Anne, and Baterina Salinas. *Philippines' Resistance: The Last Allied Stronghold in the Pacific*. Pacific Atrocities Education, 2017.

Internet Sources

Achenbach, Joel. "Did the news media, led by Walter Cronkite, lose the war in Vietnam?" *Washington Post*, (May 25, 2018), https://www.washingtonpost.com/national/did-the-news-media-led-by-walter-cronkite-lose-the-war-in-vietnam/2018/05/25/a5b3e098-495e-11e8-827e-190efaf1f1ee_story.html.

Burauen Map, https://pacificwrecks.com/airfields/philippines/san_pablo/maps/paratrooper-attack.html.

Buri Airfield, https://pacificwrecks.com/airfields/philippines/buri/1945/buri.html.

Families at War, World War II and Death, C. J. Humphrey, March 2014, https://familiesatwar2014.wordpress.com/2014/03/23/world-war-ii-and-death/.

Invasion of Leyte Map, https://pacificwrecks.com/provinces/philippines/leyte/maps/map-invasion-leyte-10-20-44.html.

Janes, Colleen. Tudie Rose Blog, https://potrackrose.wordpress.com/2013/09/20/guest-post-battle-of-buffalo-wallow-wwii/.

Japanese attack on the Burauen airfields, https://pacificwrecks.com/airfields/philippines/san_pablo/maps/paratrooper-attack.html.

Japanese map of airfields, https://pacificwrecks.com/history/nippon_news/leyte/leyte-map.html.

Japanese soldiers studying topo map for operation Te-Go, https://pacificwrecks.com/history/nippon_news/leyte/map.html.

Jump leader Shirai leading Wa, https://pacificwrecks.com/history/nippon_news/leyte/tsuneharu-shirai.html.

Kiowa painting of The Buffalo Wallow Fight, part of the Red River War, September 10th, 1874, http://www.ntxe-news.com/cgi-bin/artman/exec/view.cgi?archive=78&num=102449.

Major campaigns of the World War II Pacific Theatre of Operations, https://slideplayer.com/slide/7932976/

Melzer, Jürgen Paul. "Heavenly Soldiers and Industrial Warriors: Paratroopers and Japan's Wartime Silk Industry." *Asia-Pacific Journal* 18, Issue 17, Number 2, (September 1, 2020), Article ID 5462, https://apjjf.org/2020/17/Melzer.html.

National World War II Museum site, https://www.
nationalww2museum.org/students-teachers/student-resources/
research-starters/research-starters-us-military-numbers.

Nippon News. "Leyte Paratrooper Attack." Newsreel, Tokyo, Japan,
December 6, 1944, https://search.alexanderstreet.com/preview/
work/bibliographic_entity%7Cvideo_work%7C2019135.

Operation Te-go at airfield, https://pacificwrecks.com/history/nippon_
news/leyte/loading-wide.html.

Operation Wa paratroopers loading ammo, https://pacificwrecks.com/
history/nippon_news/leyte/load-supplies.html.

Pleasant, Keri, JMC Historian. "Honoring Black History
World War II Service to the Nation." (February
27, 2020), https://www.army.mil/article/233117/
honoring_black_history_world_war_ii_service_to_the_nation.

Rare Historical Photos. "Captain Nieves Fernandez shows to an
American soldier how she used her long knife to silently kill
Japanese soldiers during occupation – 1944."(June 21, 2014), https://
rarehistoricalphotos.com/captain-nieves-fernandez-1944/.

Southerland, Edward. "The Fight at Buffalo Wallow, September 22,
2019." http://www.ntxe-news.com/cgi-bin/artman/exec/view.
cgi?archive=78&num=102449.

Stanley, Tim. "Black History Month: Army's 'Buffalo Soldiers' help build
Fort Sill." *Tulsa World.* (February 20, 2019), https://tulsaworld.com/
news/local/black-history-month-armys-buffalo-soldiers-help-build-
fort-sill/article_11539aeb-87e8-5931-af80-60d560191235.html.

Te-Go Helen bomber, https://pacificwrecks.com/history/nippon_news/
leyte/ki49-takeoff.html.

Te-Go Loading Plane, https://pacificwrecks.com/history/nippon_news/ leyte/load-plane.html.

U.S. Army History of the Leyte Invasion, https://history.army.mil/ brochures/leyte/leyte.htm.

U.S. Army Medical Department site describing the chain of evacuation in WWII, https://www.med-dept.com/articles/ ww2-military-hospitals-general-introduction/.

U.S. Army Medical Department site describing medical units on Leyte, https://www.med-dept.com/articles/ ww2-military-hospitals-pacific-theater-of-operations/.

U.S. Army Medical Department site with field reports from Leyte, https://history.amedd.army.mil/booksdocs/wwii/actvssurgconvol2/ chapter13.3.htm.

U.S. transports unloading at Leyte, 1944, https://pacificwrecks.com/ ships/usn/LST-245/lst-245-10-20-44.html.

U.S.S. *Little* DD803 History, https://dd803.org/.

Wisconsin Veterans Museum, Madison, WI, Walter and Eda Teague Archive, https://wisvetsmuseum.pastperfectonline.com/ archive/021AF406-8F3D-4BFA-AA5E-964463363527.

For those of you who want to conduct research in a family member's World War II service, the following are great resources to guide you. In my research, I started with my Dad's Service Record which identified the unit he was in, the campaigns he participated in, and the medals received. From there I started doing Google searches to link to historical references and information from other veterans. The following two books provide tips for starting the journey.

Gawne, Jonathan. *Finding Your Father's War: A Practical Guide to Researching and Understanding Service in the World War II U.S. Army.* Havertown, PA and Oxford: Casemate, 2006.

Johnston, W. Wesley. *Dad's War: Finding and Telling Your Father's World War II Story.* Workshop Book, Wesley Johnston, 2014.

Library of Congress - Veterans History Project, http://www.loc.gov/vets/.

The website above is a great resource. If you have your relative's service record and know the unit they served in or the campaigns they fought in, you may be fortunate enough to find an interview with someone who served with them. It was here that I found interviews with Walter and Eda Teague, and Chet Gjertson, all who served with my Dad. I was thrilled to actually hear their voices and relate their stories to those that I had heard. The interviews were conducted late in their lives, so some of their recollections were a bit fuzzy. If your relative left one of these interviews, be very thankful.

As I've stated in the book's Introduction, many of the World War II veterans just didn't talk about it. You may have been with a relative, family friend, teacher, or coach and never knew their story. Here's a case in point that I'll describe. Baseball was a passion of mine growing up. I had a coach through a number of years who frankly, was very tough on us. He commanded a great deal of respect. I knew down deep, if I ever "popped off" to him, I'd probably end up going home with my teeth in my hand. But as I look back, he taught us many great values, i.e., how to practice at full intensity, how to think in situations, and how to compete. These were life lessons that served me well through college and my business career. One evening after baseball practice, while I was in high school, I was invited over to his house for dinner. His son was one of my good friends. After dinner I ended up in the den watching TV with the coach. My friend had to leave the house for another obligation. As I sat with the coach, watching a K.C. Royals game, I looked above the recliner he was sitting in. I recognized an item in a display hanging on the wall behind him as a Purple Heart medal. I got up and took a closer look. In the display was a black and white photo of a young man standing on the wing of a P-51 Mustang, a famous fighter aircraft of World War II. Being a WWII "geek" at the time, I recognized the significance of what I saw. I asked, "Coach, is this you? You flew a P-51?" I knew the answer by his look, so I continued, "How did you get the Purple Heart?" He then told me, "I was on the tail of a Japanese Zero and closing in on him. I just had the intuition to look back over my left shoulder. When I did a bullet busted through the cockpit and grazed my forehead. Another Japanese Zero had come in behind me. If I hadn't moved my head at that exact time, I would've been killed." I was stunned. This was someone I'd been around for a number of years. I had even noticed the scar across his head. But I never knew his story. The discipline he taught now made a lot of sense to me. Many others like him had their stories that went unknown. Coach Howard, you were another hero from the Greatest Generation.

The following link on YouTube, by Mark Felton Productions, provides actual footage of the Japanese paratroopers and their attack on Leyte, December 6th, 1944. It's chilling to hear the actual sounds of a "banzai charge" as members of the 44th did.

https://www.youtube.com/watch?v=2O5giMK7L9Y

Finally, please check out the "Buffalo Wallow" blog for additional stories, links, photos, and historical references. Your comments and reviews are appreciated!

http://buffalowallow.com

10174515R00118